Geology and Ore Deposits of the South Silverton Mining Area, San Juan County, Colorado

By DAVID J. VARNES

SOUTH SILVERTON MINING AREA, SAN JUAN COUNTY, COLORADO

GEOLOGICAL SURVEY PROFESSIONAL PAPER 378-A

Prepared in cooperation with the Colorado Metal Mining Fund and the Colorado Geological Survey Board

UNITED STATES GOVERNMENT PRINTING OFFICE, WASHINGTON : 1963

UNITED STATES DEPARTMENT OF THE INTERIOR

STEWART L. UDALL, *Secretary*

GEOLOGICAL SURVEY

Thomas B. Nolan, *Director*

For sale by the Superintendent of Documents, U.S. Government Printing Office
Washington 25, D.C.

CONTENTS

ILLUSTRATIONS

[All plates in pocket]

TABLES

SOUTH SILVERTON MINING AREA, SAN JUAN COUNTY, COLORADO

GEOLOGY AND ORE DEPOSITS OF THE SOUTH SILVERTON MINING AREA,
SAN JUAN COUNTY, COLORADO

By DAVID J. VARNES

ABSTRACT

The South Silverton mining area is in the western part of the San Juan Mountains, San Juan County, in southwestern Colorado. The area, which includes about 18½ square miles, has produced ores of gold, silver, lead, copper, and zinc having a gross value of more than $61 million.

The western part of the San Juan Mountains has been carved by glaciers and streams from a great domed volcanic field. Flows, tuffs, and breccias of rhyolitic, andesitic, and latitic composition having a maximum thickness of nearly 1 mile rest horizontally upon older tilted and eroded metamorphic and sedimentary rocks. The metamorphic rocks are Precambrian schist, gneiss, quartzite, and slate. The sedimentary rocks range in age from Cambrian to Late Cretaceous. Minor igneous activity occurred in Late Cretaceous or early Eocene time in some parts of the San Juan Mountains, but the great bulk of volcanic material accumulated during the middle and late parts of the Tertiary period. The volcanic rocks have been intruded by many stocks and irregular bodies of quartz monzonite, by many dikes of andesite and latite, by a few dikes of granite porphyry, and by one of rhyolite.

The Silverton region is the site of the dominant geologic feature of the western San Juan Mountains—the Silverton caldera, a roughly elliptical basin that formed in middle Tertiary time. Local areas within the basin subsided but filling by flows and small intrusions more than kept pace with subsidence. One of these areas of subsidence, about 8 by 10 miles, more or less elliptical in outline, and bounded by faults, lies at the north edge of the area covered in this report. The rocks within 2 or 3 miles of the southern periphery of the subsided block are broken by fractures that controlled ore deposition within the South Silverton mining area. The forces causing fracturing are believed to have acted radially from the subsided block.

Three systems of fractures were formed. The first is a system of unmineralized concentric fractures and dikes that are parallel to the border of the subsided block. The second, or western shear system, includes a group of mineralized northwestward-trending shear fractures and some andesite or latite dikes, a group of unmineralized complementary northeastward-trending shear fractures, and a group of very productive north-northwestward-trending tension fractures. The third, or eastern shear system, includes a group of arcuate granite porphyry dikes and a group of productive northwestward-trending veins along the eastern border of the area and east beyond the map area.

Almost all the metalliferous deposits are in veins. These range in size from a few inches wide to broad zones more than 100 feet wide, of closely spaced veins. The ores are predominantly low-grade ores of gold, silver, and base metals in minable bodies from a few tens to several thousands of feet long. Almost all the production has come from a triangular area bounded on the north by the subsided block of the caldera, on the south and west by a large northwestward-trending shear fracture, and on the east by the arcuate group of dikes.

The chief vein minerals are pyrite, galena, sphalerite, chalcopyrite, and tetrahedrite-tennantite in a gangue of quartz and minor amounts of calcite. Native gold, argentiferous galena, and argentiferous gray copper ore have accounted for most of the dollar value of products from the veins.

Small ore bodies of the replacement type are found within limestone of Mississippian and Devonian age near some of the larger veins and intrusive bodies.

INTRODUCTION

LOCATION AND GEOGRAPHY

The South Silverton mining area is immediately southeast of the town of Silverton, San Juan County, in southwestern Colorado (fig. 1). The town of

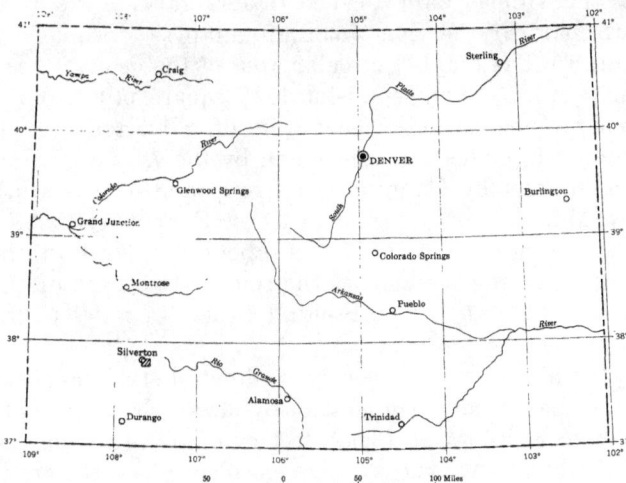

FIGURE 1.—Index map of Colorado showing location of the South Silverton mining area, San Juan County, Colo.

A1

FIGURE 2.—Bakers Park and Silverton viewed from the southwest. U.S. Highway 550 curves at the lower left; Cement Creek enters Bakers Park at the middle left and joins the Animas River at the right. The mountains in the background consist of Burns quartz latite overlain by pyroxene-quartz latite and are within the subsided block of the Silverton caldera. The ring-fault zone trends down the Animas Valley from the Shenandoah-Dives mill, at right center, toward Silverton.

Silverton itself lies in a relatively flat and open reach of the Animas Valley, called Bakers Park, in the western part of the San Juan Mountains. (See figs. 2 and 8.) The roughly circular area of the geologic map map (pl. 1) includes about 18½ square miles of the mountainous country southeast of Silverton. It is bounded on the west and north by the Animas River, on the east by Cunningham Creek, and on the south by Mountaineer Creek and Deer Park Creek. Altitudes range from 9,125 feet above sea level in the canyon of the Animas, at the southwest corner of the area, to 13,451 feet on Kendall Peak, 2¾ miles to the northeast.

Within this area nearly a dozen horn-like peaks and sharp ridges separated by deep glacial cirques rise to altitudes of 13,000 feet or more. (See figs. 3, 7, 10, and 24.) Exposures are excellent along the crests and upper flanks of the ridges, but the bedrock along the lower parts of the valley walls and floors of the cirques is largely concealed by accumulations of talus.

The timbered slopes along the south side of the Animas Valley are extensively covered with glacial moraine. Several of the high basins within the cirques hold ponds or small lakes; the largest is Silver Lake (fig. 23).

Roads skirt the northern and eastern edges of the area but none give good access into the interior. Silverton is adjacent to U.S. Highway 550, which passes over the mountains by way of Red Mountain Pass from Ouray, 24 miles to the north, to Durango, 53 miles to the south. The community is also served by the narrow-gage line of the Denver and Rio Grande Western Railroad that follows the Animas River upstream from Durango. A gravel road, State Highway 110, follows the Animas River upstream, eastward from Silverton. From this highway a side road branches off to Cunningham Gulch as far as the Highland Mary mill, and another goes up Arrastra Gulch for about 2 miles. The higher ridges east of Arrastra Basin are most easily reached by way of the Shenandoah-Dives aerial tram and mine workings. A few trails suitable for horses lead from the main valleys into the larger basins and across the intervening ridges. but much of the area is accessible only on foot. Most of the cirque heads and many of the high cliffs along Cunningham Gulch could not be examined at close range. Timberline is at about 11,500 feet.

PREVIOUS STUDIES

The geology and ore deposits of the western San Juan Mountains have been under study intermittently for the past 60 or 70 years by many geologists and mining engineers. The principal reason for the studies is their application to finding and mining the ores, but the entire region displays the results of earth deformation and volcanic activity so well that it also has long been the subject of more purely scientific investigations. The literature of geologic work bearing

directly or indirectly upon the Silverton area may be divided roughly into three groups on the basis of time, the points of division being the appearance of major studies about 1900 by Cross and Ransome, and in 1933 by Burbank.

Works published before about 1900 are mostly of a general nature; they contain occasional references to the rapidly developing mining industry of the region but are of interest principally as a record of data that can no longer be obtained elsewhere. Among these are reports and articles by F. M. Endlich (1876), T. B. Comstock (1883), S. F. Emmons (1888), and T. A. Rickard (1903).

F. L. Ransome (1901) made the first detailed study of the mineral deposits. His work was followed by the Silverton folio (Cross, Howe, and Ransome, 1905), in which the results of several years of geologic mapping were combined with a summary of the mineral de-

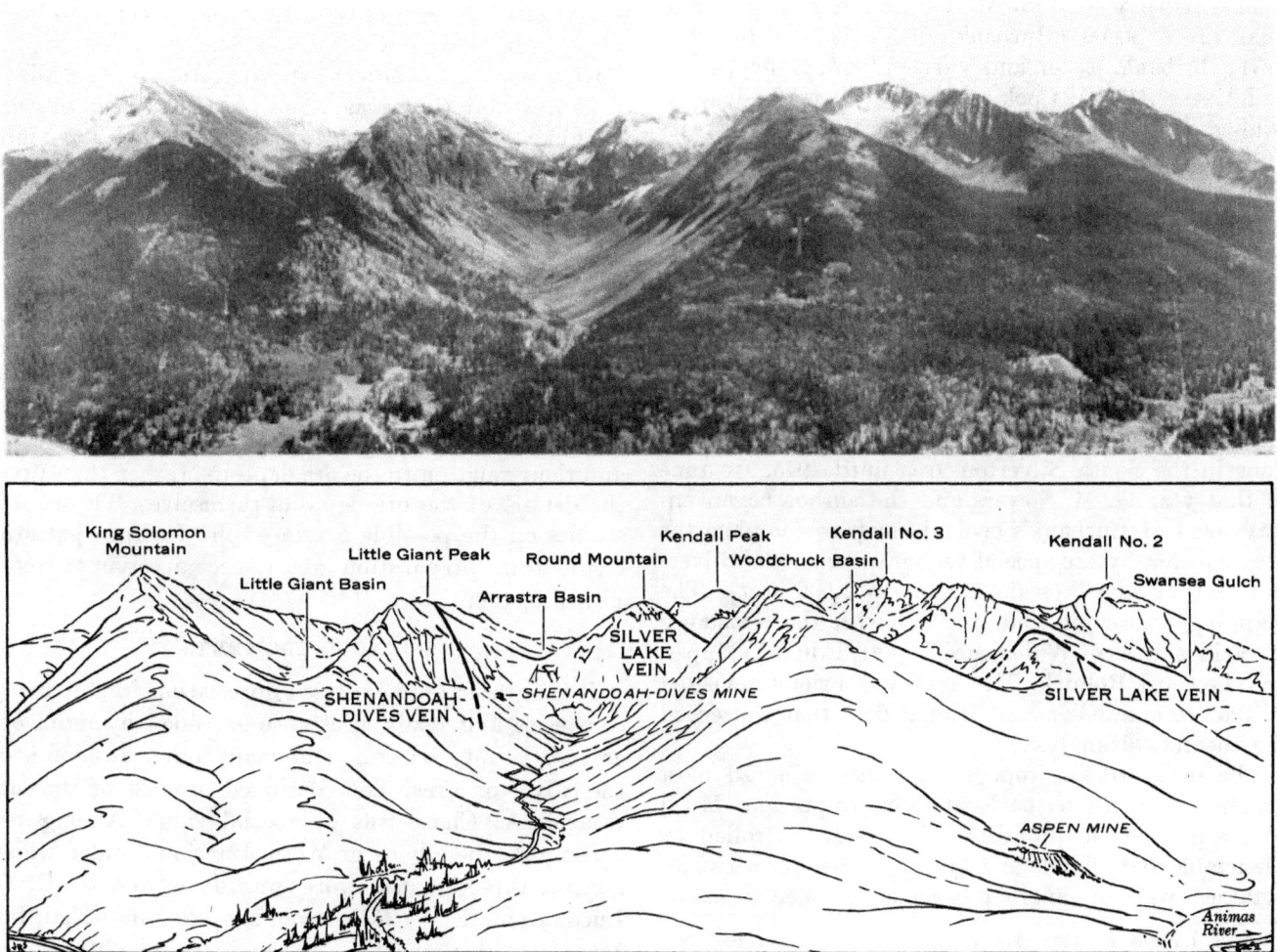

FIGURE 3.—Sketch and view southward of hornlike peaks, sharp ridges, and deep glacial cirques in the northern part of the South Silverton mining area from a point above the Shenandoah-Dives mill on the north side of the Animas Valley. The timbered south slope of the Animas Valley is in the foreground.

posits. About 1905 there began a period of transition in which shallow mining and shipping of crude ore slackened and operations were consolidated into fewer and larger mines with facilities for milling large tonnages of ore. This trend continued for many years, and accounts of the development of the larger mines of the district may be found in papers by W. C. Prosser (1914) and C. A. Chase (1929). An excellent summary of the mining activity is contained in C. W. Henderson's Mining in Colorado (1926).

Geologic mapping of this mining district at a scale of 1 inch to 1,000 feet was started in Arrastra Basin in 1932 by W. S. Burbank. His report and progress map at a scale of about 1 inch to 2,000 feet covered the more heavily mineralized area of Arrastra Basin and vicinity (Burbank, 1933a) and marked the beginning of some correlation between regional structure and the vein systems in the heavily mineralized sector southeast of Silverton.

This paper was followed by others referring to San Juan ore deposits (Burbank, 1933b, 1935, 1940, 1941, 1951; Burbank, Eckel, and Varnes, 1947; Kelley, 1946; and Varnes, 1948). Cook[1] and Hagen[2] made detailed studies of some of the ore deposits that lie just east of Cunningham Gulch. Their work is referred to in the sections on Tertiary structure and ore deposits. Several articles on Silverton have appeared in the yearbooks of the Colorado Mining Association. The geology of the whole San Juan region was summarized by Cross and Larsen (1935) and more recently described in a comprehensive study (Larsen and Cross, 1956).

FIELDWORK

After Burbank's early work no further mapping was done in the South Silverton area until 1945. In June of that year G. M. Sowers and the author began enlargement of Burbank's geologic map to complete the area covered by the special topographic sheet "Silverton and vicinity" (scale 1 inch to 1,000 feet). The mapping was completed in 1946 with the assistance at various times of Waldemere Bejnar, Eduardo Mapes, and Leonard Rolnick. The area was briefly revisited in 1952 to obtain some additional data that concerned the theoretical analysis.

The detail of the topographic sheet allowed most geologic features to be located by inspection. Much of the geologic mapping was, however, controlled by planetable and alidade and by supplementary altitudes obtained with an aneroid barometer. Some locations

were made by compass bearings and resection; but the compass needle occasionally was deflected as much as 14° by the magnetic properties of the volcanic rocks. Many of the less important features, such as the boundaries of talus and other surficial deposits, were mapped with the aid of a graphic locator (Varnes, 1946). This technique was developed and tested during the mapping of this area.

The aerial photographs available at the time of mapping were unsuited to geologic work, but they did aid in planning traverses. Excellent aerial photographs were taken in 1951. These and the multiplex topographic sheets prepared from them were studied closely, but they appear to require no significant revision of the 1931 topographic base or of the geologic mapping.

PURPOSE OF THE WORK

The purpose of the Geological Survey's program of mapping parts of the western San Juan at a scale of 1 inch to 1,000 feet has been first, to present a geologic framework within which more detailed data may be fitted; second, to indicate the pattern of mineralized fractures and the areas where successful exploration is unlikely; and third, to furnish basic data and maps of a convenient scale for future planning and development. The more specific objectives of mapping southeast of Silverton were to extend knowledge of the rock types, the dike and vein systems, and the mineral deposits around the southern periphery of the Silverton caldera. As mapping progressed beyond Arrastra Basin, it became apparent that the major geologic problem was the structural control of ore deposition. The main emphasis therefore has been placed upon study of the fracture patterns of the area, their origin and their relation to the ore deposits, rather than upon the details of the ore deposits themselves. Theoretical studies on the possible origin of the fracture patterns resulted in preparation of an accompanying report (Varnes, 1962).

ACKNOWLEDGMENTS

It is a pleasure to express appreciation for the many courtesies and ready assistance extended by mine operators, county officials, and many other residents of the Silverton area. The continued interest of the late Charles A. Chase was of special value. Among my colleagues, the advice of W. S. Burbank, under whose general direction the work was done, and of R. G. Luedke and E. C. Robertson has been most helpful. Mr. Burbank's reports on the regional structure of the Telluride and Silverton districts (1933a, 1941) provided a firm basis for an understanding of the present

[1] Cook, Douglas R., 1952, The geology of the Pride of the West vein system: Ph.D. thesis (no. 747), Colorado School of Mines.
[2] Hagen, John C., 1951, The geology of the Green Mountain mine, San Juan County, Colorado: M.A. thesis (no. 727), Colorado School of Mines.

and future geologic work, and his report on Arrastra Basin (1933a) presents a geologic picture that may be extended with only minor additions or changes to the surrounding area covered by this more recent work. Repetition of some of the general data obtained by Burbank and others is inevitable, but many of the details of the Arrastra Basin area already published are not repeated here.

This study, like the earlier work by Burbank, was made with financial cooperation from the Colorado State Geological Survey Board and the Colorado Metal Mining Fund.

OUTLINE OF GEOLOGIC HISTORY

The rocks of the western San Juan Mountains present a long and fairly complete record of the geologic events that have affected this region from very ancient to recent geologic time. Within any small area, such as that covered in this report, only fragments of the record may remain or be exposed for observation. So the following brief outline of geologic history necessarily is derived from previous studies by other geologists over the greater part of southwestern Colorado. The relation of the South Silverton mining area to the whole is then summarized.

Precambrian rocks of great variety and complex structure occur in the San Juan Mountains. They may be divided into four major units, which are from youngest to oldest: granitic rocks, the Needle Mountains group (mostly quartzite and conglomerate), the Irving greenstone, and ancient schist and gneiss. Several stages of mountain building, intrusive and extrusive igneous activity, and erosion are recorded during this period. At the end, a long period of erosion extending into Cambrian time wore the land to a smooth plain that was later submerged beneath the sea.

A thin layer of gravel and sand, now hardened into Ignacio quartzite of Late Cambrian age, accumulated on the beveled surface of the Precambrian rocks. This was followed by a succession of other Paleozoic strata, including Upper Devonian shale and limestone; Mississippian marine limestone; Mississippian and Pennsylvanian shale, limestone, and sandstone; and Permian sandstone, shale, and limestone. This succession of Paleozoic strata is without angular unconformity, but there are several disconformities and gaps in the sedimentary sequence. Cambrian rocks are absent in the far eastern part of the San Juan Mountains, where Ordovician limestones and sandstones lie directly on the Precambrian rocks. In the western part of the San Juan Mountains, Ordovician, Silurian, and Lower and Middle Devonian beds are missing. A period of uplift

and erosion that followed deposition of the Mississippian Leadville limestone is generally represented in the western part of the San Juans Mountains by the deeply weathered karstlike upper surface of this limestone. Upon this surface lies the reddish residuum and shale of the Molas formation of Pennsylvanian age.

Continental deposition continued into the Mesozoic era with apparent regularity over most of this area. At Ouray, however, the Upper Triassic Dolores formation lies in angular unconformity upon inclined strata of Permian and older age, indicating, according to Burbank (1940, p. 197), an ancestral (late Paleozoic) domal uplift of the San Juan Mountains with local subordinate radial axes of uplift in the Uncompahgre district. A later period of intense local crumpling and extensive erosion before deposition of the Entrada sandstone and Morrison formation of Late Jurassic age is recorded in the eastern part of the San Juan Mountains. Elsewhere the sequence is without this angular unconformity, and, where exposed, it continues regularly through a thick series of Cretaceous terrestrial and marine sandstone and shale beds. Altogether, about 9,000 feet of Mesozoic strata were deposited over most or all of the San Juan region.

The San Juan region was again uplifted near the end of the Cretaceous period into a great dome and subjected to deep erosion. At the end of the Cretaceous period or in early Tertiary time began the first of the many periods of intrusive and volcanic activity that characterize the Tertiary history of the San Juan Mountains. This initial period of intrusion produced the dikes, sills, laccoliths, and small plugs of granodiorite porphyry near Ouray, together with genetically related ore deposits.

After a period of structural unrest, the region was again eroded to a surface of low relief. Sand and gravel of variable thickness and composition were deposited on parts of this surface. In the western part of the San Juan region this indurated detritus is called the Telluride conglomerate of Oligocene(?) age, and the surface on which it rests is known as the Telluride erosion surface. Another considerable period of erosion may have occurred between the deposition of the Telluride conglomerate and the deposition of the San Juan tuff. The San Juan tuff, of Miocene(?) age, overlies the Telluride conglomerate without angular unconformity where the latter is present, and it rests on a very irregular surface of pre-Tertiary rocks where the conglomerate is absent. The tuff attains a maximum thickness of nearly 3,000 feet near Ouray. The San Juan tuff was followed, after an interval of erosion, by a complex sequence of lavas, tuffs, and agglomerates, known as the Silverton volcanic series of Miocene age, which are thickest within a basin-shaped volcanic depression known as the Silverton caldera. Volcanic activity, represented by the Potosi volcanic series (middle or late Tertiary) and younger formations, continued with intervals of erosion, in various parts of the San Juan Mountains through Miocene, Pliocene, Pleistocene, and possibly into Recent time. The result was a great dome or plateau of volcanic rocks over 100 miles across and more than a mile thick.

Faulting, fissuring, and intrusive activity, followed by formation of the major ore deposits of the San Juan mining region, occurred in late Tertiary time— Miocene or later—for many of the veins and intrusive rocks cut rocks of the Potosi volcanic series.

One of the periods of erosion during late Tertiary time was of sufficient duration to reduce the whole region to a surface of low relief, the San Juan peneplain of Pliocene(?) age. A few mountains 1,000 feet or more in height rose above this gently rolling surface. The surface was partly covered by debris from extensive volcanic eruptions, the Hinsdale formation of Pliocene(?) age, and then again domed, faulted, and gently tilted to the east during an uplift that marked the transition from Tertiary to Quaternary time. This uplift took place in two main stages. The first stage was followed by erosion to a mature topography, the Florida erosion surface, and later by the Cerro stage of glaciation. The Cerro glacial stage was followed by a second stage of uplift, the Canyon cycle of erosion, and the Durango glacial stage. The Durango glacial stage was succeeded, in turn, by slight doming and by the third and last glacial stage referred to by Atwood and Mather (1932) as the Wisconsin. The present topographic form of the San Juan Mountains is the result of these several cycles of uplift, and erosion by ice and water during the Pleistocene. Since the retreat of the Wisconsin glaciers, some streams have renewed cutting and have deepened their valleys as much as a few hundred feet. For a full discussion of the physiographic history of the San Juan Mountains the reader is referred to the work of Atwood and Mather (1932). The glacial chronology has been revised by Richmond (1954) who refers the Cerro, Durango, and Wisconsin glacial deposits of Atwood and Mather respectively to the pre-Wisconsin, early Wisconsin (Iowan and Tazewell), and late Wisconsin (Cary and Mankato).

Silverton lies in the west-central part of the San Juan Mountains, and its geologic setting is the combined result of most of the geologic events mentioned previously. Precambrian gneiss and granite are exposed in the Animas Canyon south of Silverton (fig. 4). The Needle Mountains group and the Irving greenstone, however, crop out in the Needle Mountains, a few miles south of the area. Paleozoic sedimentary rocks are represented by the Ignacio quartzite, Elbert formation, Ouray and Leadville limestones, and by a few feet of Hermosa formation. In the southwestern part of the South Silverton area, the Tertiary volcanic rocks overlie the relatively thin sequence of Paleozoic rocks with angular unconformity; elsewhere in this area the volcanic rock rest upon gneiss and schist. The late Paleozoic deformation recorded at Ouray may have extended into this area and contributed to the steep dip of the Paleozoic rocks. Certainly the area southeast of Silverton was involved in the broad doming at the end of the Cretaceous. In fact, the mapped area may be on the north flank of a somewhat sharper subsidiary uplift having its axis in the present site of the Needle Mountains. When the erosion that produced the Telluride surface removed tilted Mesozoic and upper Paleozoic rocks from this area, the remaining lower Paleozoic rocks were beveled to a wedge that thins to the south. Evidence of late Mesozoic or early Tertiary intrusive activity and mineralization, such as found at Ouray, has not been recognized in the Silverton area.

The Telluride conglomerate was not found during the present mapping and is presumed to have been removed by erosion preceding or concurrent with eruptions of the San Juan tuff. San Juan tuff is present, although its thickness varies greatly and is small relative to that at Ouray. The wide range in thickness is due, in part at least, to erosion that followed deposition of the San Juan tuff and preceded eruption of the Silverton volcanic series. The lowest and highest units of the Silverton volcanic series are missing, but the middle part is well represented by the flows and agglomerates that form the great body of the mountains in the mapped area.

Late Tertiary structural events, including faulting, fissuring, and a type of large-scale subsidence peculiar to a few areas of the western San Juan Mountains, exerted strong control over the localization of ore deposits in the South Silverton area. Dikes and stocks of late Tertiary andesite, latite, granite porphyry, and quartz monzonite are common. Extrusive rocks of the Potosi volcanic series and younger formations are now lacking although they may once have covered most of the area.

Finally, erosion by glaciers and running water during the Pleistocene was particularly effective at Silverton, producing a valley wide enough and flat enough to contain the town, yet surrounding it with mountains of unusual grandeur.

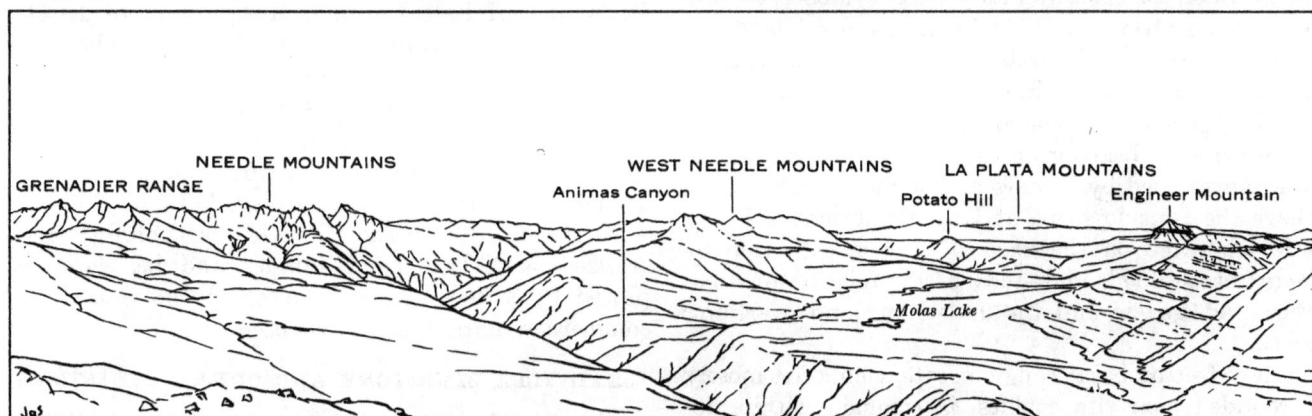

FIGURE 4.—Sketch and view southward of the V-shaped canyon of the Animas cut into Precambrian gneiss and granite. The flat area around Molas Lake at right center is thinly mantled by lower Paleozoic sedimentary rocks. Tilted Pennsylvanian and Permian strata crop out on slopes that rise to the right. The flat-topped ridge at the right is capped by flat-lying Telluride conglomerate and San Juan tuff. Engineer Mountain is on the skyline at right and the La Plata Mountains appear in the distance at right center. The Needle Mountains are in the distance on the left. Taken with phototheodolite from Kendall No. 2 Peak.

PRECAMBRIAN ROCKS

The oldest rocks in this area are Precambrian schist and gneiss. They are well exposed in the canyon of the Animas River below Silverton. They also crop out along the southern border of the area, and in the upper part of Cunningham Gulch in the southeastern part of the area. (See fig. 5.)

Banding is nearly everywhere a conspicuous feature of the Precambrian rocks. It is due to regular variation in the relative proportion of light minerals (feldspar and quartz) to the dark minerals (biotite and hornblende). The large bands, generally several feet to many tens of feet wide, probably reflect differences in composition of the original rocks. Small bands and foliae 1 mm or less wide probably resulted from recrystallization of the original rocks under high heat and pressure. Crushing and recrystallization have removed nearly all traces of original structure from the schist and gneiss, but from the overall banded appearance and general composition Cross and Howe (Cross, Howe, and Ransome, 1905, p. 3) inferred an original

assemblage of sedimentary, volcanic, and sheetlike intrusive rocks.

In the extreme southwest corner of the area, dikes and irregular masses of biotite granite have cut and assimilated the schist, and a few diabase dikes were injected along planes of foliation. Both the granite and the diabase were considered to be Precambrian by Cross and Howe (Cross, Howe, and Ransome, 1905, p. 6); the diabase is the younger judging from the mapped relations of the two rocks near the southern edge of the Silverton quadrangle. Because these intrusive rocks are relatively small and generally poorly exposed, they are not shown separately from the schist on plate 1.

The great section of Precambrian quartzite and slate of the Needle Mountains group, which is younger than the ancient schist and gneiss and probably younger than the biotite granite, does not crop out between Whitehead Gulch, 2 miles south of the southern border of the area shown on plate 1, and Ironton Park, 8 miles north of Silverton. Near Silverton, therefore, the schist and gneiss are directly overlain by undisturbed Cambrian or other Paleozoic strata, or by the Tertiary volcanic rocks.

PALEOZOIC SEDIMENTARY ROCKS

IGNACIO QUARTZITE

Paleozoic sedimentary rocks crop out in the southwestern part of the area, between Precambrian schist and gneiss and the Tertiary rocks, and in a few places near the eastern edge of the area. The oldest Paleozoic rock unit within the region is the Ignacio quartzite. The Ignacio has been long referred to the Upper Cambrian on the basis of scanty fossil evidence, *Obolus* sp.(?) from Overlook Point in the Needle Mountains (Cross, Howe, and Ransome, 1905). This interpretation was questioned by Barnes (1954), who suggested that the Ignacio is Devonian in age.

The evidence bearing on the age of the Ignacio has been re-examined by Rhodes and Fisher (1957). They believe the Ignacio to be of Late Cambrian or Early Ordovician age on the basis of a new collection of *Obolus* brachiopods and study of the relationships between the Ignacio and the underlying and overlying rocks.

A few feet of basal conglomerate, composed mostly of rounded quartzite cobbles, is common within depressions on the surface of the Precambrian rocks. The overlying strata are composed of well-bedded dense white and pink quartzite with thin partings of yellow and red shale. One bed of dense blue-gray limestone a few inches thick was seen in the Ignacio quartzite on the east side of the Animas Canyon below the trail into Kendall Gulch. The quartzite has a maximum thickness of about 100 feet, but is more commonly 50 feet or less in thickness.

ELBERT FORMATION

The Elbert formation of Late Devonian age is exposed only at the south edge of the landslide area south of the Highland Mary mine, and along the lower part of Kendall Gulch on the north side. At Kendall Gulch it consists of a few feet of yellow, green, red, and purple calcareous shale and thinly bedded limestone. At the small exposure near the landslide, the Elbert formation rests directly upon Precambrian schist; at Kendall Gulch it rests upon Ignacio quartzite. The erosional unconformity at the base of the Elbert, representing partial to complete removal of the Ignacio quartzite before deposition of the Elbert, has also been observed to the north of this area, near Ouray (Burbank, 1930, p. 157–158) and to the south, near the Needle Mountains (Cross and Larsen, 1935, p. 32).

The weak rocks that make up the Elbert formation do not tend to form good outcrops. Possibly some outcrops were not recognized, but no certain trace of the Elbert was found outside the two areas mentioned, and the formation is believed to be absent from most of the map area. In this respect, the geologic map for this report differs from that of the Silverton folio (Cross, Howe, and Ransome, 1905), which shows the Elbert formation beneath the Ouray limestone wherever the latter is exposed in the southern part of the Silverton quadrangle. The Elbert presumably was almost completely removed from this area by erosion prior to the deposition of the overlying Ouray limestone, also of Late Devonian age. For more detailed information regarding the age, lithologic character, and distribution of the Elbert formation the reader is referred to the original description by Cross (1904), and dating by Eastman (1904), to the Engineer Mountain folio (Cross and Hole, 1910), and to a general discussion of the Devonian in Colorado by Kirk (1931).

At Kendall Gulch the Elbert shale is slightly mineralized with base-metal sulfides and has been explored by a few prospect holes, but no deposits of commercial importance have been found.

LEADVILLE LIMESTONE AND OURAY LIMESTONE

The Elbert formation, where present, is overlain without angular discordance by the Ouray limestone of Late Devonian age; the Ouray limestone is overlain conformably by the Leadville limestone of Mississippian age. Where unaltered by igneous activity or vein solutions, both limestones are fine to medium grained, rudely bedded to massive and gray, except for a few pinkish beds near the base of the Ouray. In some places they may be separable in age by detailed study of the fossils that they contain, but since no lithologic distinction can be drawn in the field between the two similar-appearing limestones, they were mapped as a single unit. In the northwestern part of the area, the limestone is about 225 feet thick, but it thins southward beneath the prevolcanic erosion surface and is absent south of Deer Park Creek.

The Leadville and Ouray limestones contain two metalliferous deposits of the replacement type. One is at the Marcella mine at the far west edge of the map area and the other is at the March crosscut of the Osceola mine just beyond the eastern border of the map area. These are briefly described in the section on replacement deposits (p. A48).

PENNSYLVANIAN SYSTEM

The red shale and sandstone characteristic of the Molas formation have not been recognized within the area studied, although they are conspicuous along the road to Durango a few miles to the west. The remainder of the Pennsylvanian section, although about 2,000 feet thick a few miles west from the Animas River, is represented east of the Animas River by no more than 25 feet of green sandstone overlying Ouray and Leadville limestones along the trail between Kendall Gulch and Deer Park Creek. This sandstone is tentatively assigned to the Hermosa formation, which is well exposed west of the Animas Canyon.

TERTIARY BEDDED ROCKS

TELLURIDE CONGLOMERATE

The sequence of Tertiary bedded rocks in the western part of the San Juan Mountains begins at most places with the Telluride conglomerate of Oligocene(?) age. The conglomerate rests on an extensive, nearly smooth erosion surface developed in Eocene(?) time upon all older rocks. The geologic maps of the Silverton folio (Cross, Howe, and Ransome, 1905) show the conglomerate on the Sultan and Grand Turk Mountains about 1 mile west of the western border of the South Silverton area (see also fig. 4). The folio maps show that the Telluride is missing from its usual position at the base of the volcanic rocks along the north side of Deer Park Creek and near the mouth of Kendall Gulch but the folio maps show it 2 or 3 miles to the east at the headwaters of Mountaineer Creek. The geologic map (pl. 1) in this report differs from those of the folio by showing no Telluride conglomerate in the Mountaineer Creek area, and indeed, none at all within the area covered by the folio. This difference results from some volcanic detritus found among the conglomerate beds near the head of Mountaineer Creek. In conformance with Burbank's usage (1933a, p. 139) that the Telluride conglomerate "is of non-volcanic character and composed entirely of pre-Tertiary or early Tertiary rocks," the conglomerate beds were grouped with the overlying volcanic conglomerate in the lower part of the San Juan tuff.

SAN JUAN TUFF

The first accumulation of volcanic rocks during the middle part of the Tertiary period is represented by the San Juan tuff, comprising water- and air-laid breccia, agglomerate, and tuff. This formation extends over most of the northwestern part of the San Juan Mountains and has a maximum thickness of nearly 3,000 feet at Ouray. The San Juan tuff is exposed within the South Silverton area in the headwaters of Deer Park Creek and in the valley of Mountaineer Creek, where it ranges from 10 feet to several hundred feet in thickness.

The lower 10 to 100 feet of the formation is generally a conglomerate composed of cobbles and boulders of volcanic rocks, together with granitic and other pre-Tertiary rocks. The basal few feet of the conglomerate is locally a breccia consisting almost entirely of angular fragments of schist, gneiss, and granite. Most cobbles in the upper part of the conglomerate are of andesite and latite, and are rounded, indicating deposition from running water.

The upper part of the San Juan tuff consists of rudely bedded tuffs and agglomerate containing fine-grained to porphyritic andesite and latite fragments ranging in size from a fraction of an inch to several feet cemented in a matrix of fine tuff. The fragments are not bombs or lapilli and show no sign of having been ejected in a semiplastic condition. Some of the material may be air laid and some the product of subaerial erosion and transportation from distant volcanic accumulations. Burbank (1930, p. 189) states: "Some of the angular and more homogeneous breccias give the impression that they may have been extensive mud-flows, in which there was less opportunity for the abrasion and rounding of individual fragments."

Cross, Howe, and Ransome (1905, p. 7) state:

The colors of the formation when freshest are neutral blues, grays, and brownish-purples, with sometimes vivid greens and robin's-egg blues * * * in many places where there has been impregnation by iron sulphides subsequent oxidation has stained the tuffs red or yellow, and the decomposing action of mineral-bearing waters has often so changed the whole character of the rocks as to practically destroy all individuality.

Blue green and brownish purple are the dominant rock colors in the South Silverton area.

There are no direct means of dating the San Juan tuff. It is older than the Silverton volcanic series of Miocene age, younger than the Telluride conglomerate of possible Oligocene age, and is currently considered to be of Miocene(?) age.

The present mapping has omitted the small area of Telluride conglomerate shown on the Silverton folio map near Mountaineer Creek. Plate 1 shows instead San Juan tuff, which is here conglomeratic and rests directly upon Precambrian rocks. This relation, together with the considerable range in altitude of the base of the San Juan tuff, is evidence of erosion that followed deposition of the Telluride conglomerate and preceded or perhaps accompanied deposition of the tuff. Cross and Howe first suggested this interval of erosion (Cross, Howe, and Ransome, 1905, p. 21), but its existence, at least as based on evidence in Arrastra Basin, was later questioned by Burbank (1933a, p. 140). I agree with Burbank in assigning the breccia of Arrastra Basin, which was mapped as San Juan tuff by Cross and Howe, to erosional detritus within the Eureka rhyolite and likewise include the breccias on the west slope of Kendall Mountain within the Eureka. At Kendall Mountain, however, the relations are very complex. The rhyolitic breccia is so altered by its proximity to both the quartz-monzonite stock and the rhyolite, which may be intrusive, that the question of whether it belongs to San Juan tuff or the Eureka rhyolite is still unanswered.

SILVERTON VOLCANIC SERIES

The San Juan tuff was extensively eroded prior to eruptions of the overlying Silverton volcanic series. Evidence of this erosion, which produced a land surface nearly as rugged as that of the area today, is clearly shown by the great range in altitude of the base of the Silverton series, local nearly vertical contacts with the underlying San Juan tuff, and the total lack, in many places, of any San Juan tuff between flows of the Silverton series and pre-Tertiary rocks.

The Silverton volcanic series is made up of five formations. These are from base to top: the Picayune quartz latite, Eureka rhyolite, Burns quartz latite, pyroxene-quartz latite, and Henson tuff.

These names for the formations within the Silverton series are those adopted by Larsen and Cross (1956); the first, third, and fourth are modified somewhat from the earlier usage. Although the lithologic classification still may not be entirely satisfactory, the number of good analyses available from the Silverton area is at present not sufficient to warrant precise terminology. The series attains its maximum thickness of more than 3,000 feet within the Silverton quadrangle and extends, with diminished thickness, to the west, north, and east over an area of about 1,000 square miles. In both area and volume the Silverton series is smaller than the older San Juan tuff and the younger Potosi volcanic series. Near Silverton, however, nearly all the exposed volcanic rocks belong to the Silverton series.

Only the second, third, and fourth formations of the Silverton series have been recognized in the South Silverton area. The maximum aggregate thickness of the Eureka rhyolite, Burns quartz latite, and pyroxene-quartz latite is at least 3,000 feet and may be as much as 4,000 feet. The thicknesses of the Eureka rhyolite and Burns quartz latite are variable because of the irregularity of the surfaces upon which they were extruded. In general, the thickness of the volcanic flows decreases to the south as the underlying eroded surface of San Juan tuff and pre-Tertiary rocks rises toward a highland area.

Chemical analyses of rocks from the Silverton volcanic series collected within the area and from nearby are given in tables 1, 2, and 3.

TABLE 1.—*Analyses, norms, and modes of Eureka rhyolite and similar average igneous rocks*

[Analyses for columns 1-3 were made in the rapid-rock-analyses laboratory, U.S. Geological Survey, under the supervision of W. W. Brannock. Norms for samples in columns 1-3 were computed from a modification of the chemical analyses in which the percentages of oxides were recomputed on a basis of 0.6 percent H_2O and 0.0 percent CO_2 to remove some of the effect of alteration so that a comparison could be made more easily between these norms and the average norms of fresh rocks reported by Nockolds]

	1	2	3	4	5	6
Analyses						
SiO_2	64.9	52.7	52.6	65.88	66.27	73.66
Al_2O_3	17.6	17.2	15.9	15.07	15.39	13.45
Fe_2O_3	1.8	3.9	5.2	1.74	2.14	1.25
FeO	1.6	5.9	4.9	2.73	2.23	.75
MgO	1.3	6.0	4.5	1.38	1.57	.32
CaO	3.1	2.0	8.4	3.36	3.68	1.13
Na_2O	3.8	5.0	2.5	3.53	4.13	2.99
K_2O	4.1	1.8	1.9	4.64	3.01	5.35
H_2O-	} 1.3	3.8	2.2			
H_2O+				.52	.68	.78
TiO_2	.49	1.0	1.0	.81	.66	.22
P_2O_5	.18	.40	.38	.26	.17	.07
MnO	.10	.18	.15	.08	.07	.03
ZrO_2						
CO_2	.07	.08	.72			
S						
BaO						
SrO						
Li_2O						
FeS_2						
Total	100	100	100			
Norms						
Q	19.0	1.9	8.5	18.8	20.8	33.2
or	24.5	11.1	11.1	27.2	17.8	31.7
ab	32.0	44.0	22.0	29.9	35.1	25.1
an	14.5	7.8	27.0	11.7	14.5	5.0
C	1.7	4.4				.9
di						
hy						
$CaSiO_3$			5.5	1.4	1.3	
$MgSiO_3$	3.2	15.5	11.5	3.4	3.9	.8
$FeSiO_3$	1.0	6.3	3.4	2.4	1.3	
mt	2.6	6.0	7.7	2.6	3.0	1.9
il	.9	2.0	2.0	1.5	1.4	.5
hm						
ap	.3	1.0	1.0	.6	.3	.2
Modes						
Quartz	<5		Trace			
Orthoclase	15					
Plagioclase	10(an$_{10-45}$)		29(an$_{45-65}$)			
Hornblende	Relict					
Pyroxene	Trace		9			
Biotite						
Magnetite	Trace		1			
Chlorite						
Apatite	Trace					
Groundmass	70		61			

1. Eureka rhyolite. Quartz latite welded tuff (46DV1), from Swansea Gulch near the Scranton City mine, 37°48'08'' N., 107° 38'24'' W. Lab. No. 151412.
2. Eureka rhyolite. Flow-breccia facies (46DV32) from southwest slope of Kendall No. 2 Peak, about 37°47'00'' N., 107°39'10'' W. Lab. No. 151419.
3. Eureka rhyolite(?). Andesitic facies (46DV19a), from a point near head of Spencer Basin, 37°46'21'' N., 107°36'05'' W. Only rock in the area that contains fresh pyroxene. May be equivalent to the Picayune quartz latite. Lab. No. 151418.
4. Hornblende-biotite adamellite, average of 41 analyses (Nockolds, 1954).
5. Rhyodacite and rhyodacite-obsidian, average of 115 analyses (Nockolds, 1954).
6. Calc-alkali rhyolite and rhyolite-obsidian, average of 22 analyses (Nockolds, 1954).

EUREKA RHYOLITE

The Eureka rhyolite and overlying Burns quartz latite have been described in considerable detail by Burbank (1933a, p. 141–145), who tentatively divided the Eureka in the Arrastra Basin area into three members: A lower rhyolite flow, a medial tuff breccia, and an upper flow breccia. These divisions are not recognizable throughout the surrounding area, and for this reason I did not map the three members of the Eureka rhyolite separately.

The lower parts of the Eureka rhyolite are irregular and, in part, interbedded with breccia and tuff derived from the previous land surface. The early flows accumulated in valleys that had been eroded in San Juan tuff and older rocks. The valleys were finally filled and the upper flows of rhyolite and rhyolitic flow breccia spread out nearly horizontally on a surface of low relief.

At the head of Spencer Basin one of the flows not far above the San Juan tuff is a relatively fresh dark-gray pyroxene andesite. Stubby plagioclase phenocrysts, 1 to 3 mm in length, make up one-quarter to one-third of the rock. The plagioclase is zoned and ranges in composition from An_{45} to An_{65}, which is considerably more calcic than the albite or oligoclase generally found in rocks of the Eureka rhyolite. Pyroxene is present to the extent of a few percent and is relatively unaltered. This rock may possibly belong more properly with the Picayune quartz latite of Larsen

and Cross (1956). Analyses of this rock and of more typical specimens of Eureka rhyolite are given in table 1.

Nockolds' averages are given for comparison with column 1, which represent the common facies of Eureka rhyolite more closely than those of columns 2 and 3. The normative amounts of Q, or, ab, and an in column 1 fall within the range of norms between average hornblende-biotite adamellite and average rhyodacite-obsidian, but are quite different from the norm of average calc-alkali rhyolite and rhyolite-obsidian. The analyses support the general impression gained in the field that the Eureka rhyolite, while greatly variable in composition, is in general less silicic and more aluminous than the name would suggest.

The Eureka rhyolite is well exposed on the west side of Cunningham Gulch, as shown in figures 5 and 6. In the slopes below Dives Basin, the three members (not mapped separately) can be distinguished as: (1)

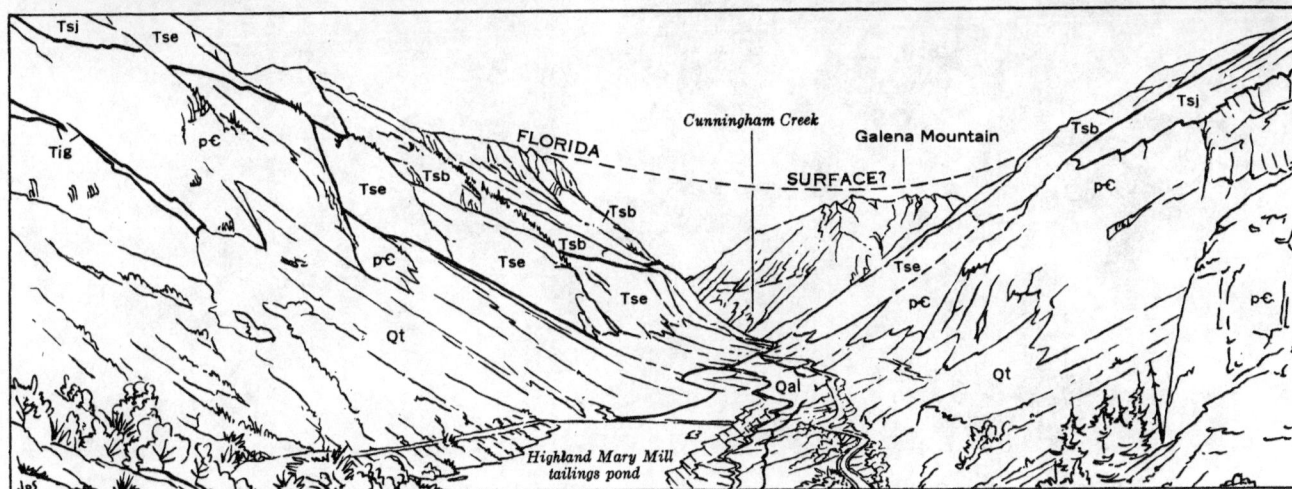

FIGURE 5.—Sketch and view northward down Cunningham Gulch. Note that the upper surface of the Precambrian gneiss declines toward the north; granite porphyry dikes are on the left. Qal, alluvium; Qt, talus; Tig, granite porphyry; Tsb, Burns quartz latite, undivided; Tse, Eureka rhyolite; Tsj, San Juan tuff; p€, Precambrian schist and gneiss.

FIGURE 6.—West side of Cunningham Gulch viewed northwestward from the Green Mountain mine. The main cliff is composed of Eureka rhyolite. The base of the Burns quartz latite is near the top of the cliff at the left. The irregular white mass to the right of the dark outcrop at the base of the cliff is Paleozoic limestone that apparently was caught and incorporated into the flow breccia.

about 650 feet of light-gray flow-banded fine-grained rhyolite flows or welded tuff overlain by (2) about 70 feet of purple to pink tuff breccia overlain by (3) about 60 feet of flow breccia. On the west slope of Kendall Peak No. 2, 1½ miles south of Silverton, the total thickness of the Eureka is about 2,000 feet. The interval in the lower 600 to 900 feet consists of dense blue-gray porphyritic rhyolite and latite. Some of this interval may be intrusive, as suggested by the steep flow lines and irregular contacts of the rhyolite with lenses of dark greatly altered breccia similar to the San Juan tuff. The remainder of the formation includes light-gray rhyolitic and quartz latitic flows and flow breccia. The top of the Eureka rhyolite in the western part of the South Silverton area is marked in many places by a platy tuff ranging in thickness from a few feet to 100 feet. Generally this tuff has the texture of very coarse sandstone and is given a mottled appearance by numerous green platy fragments in a matrix of white feldspar grains. Locally, the tuff is very fine grained, dense, and lavender or light green.

Much of the material in the Eureka formation that has been called rhyolite flow may be welded and recrystallized tuff. A specimen taken just above the Scranton City mine, 1.4 miles southeast of Silverton, is typical of the massive light-gray lower part of the Eureka. To the eye it appears wholly crystalline with many gray or greenish stubby feldspar phenocrysts, 3 to 4 mm long, in a fine-grained light-gray matrix. Angular fragments of a generally darker color are common, and the rock as a whole shows an alinement of the long dimensions of crystals and rock fragments. In thin section, the fragmental character of the rock is well shown. Orthoclase and plagioclase (albite to andesine) phenocrysts make up about 30 percent of the rock. The remainder is a finely crystalline to glassy groundmass that does not show flow structure. The feldspar phenocrysts appear to be fragments of larger crystals and many show corroded and embayed borders. Fragments of exotic material are common.

BURNS QUARTZ LATITE

The Burns quartz latite overlies the Eureka rhyolite with erosional unconformity, and consists of breccias, flows and tuffs of dominantly latitic and andesitic composition having a maximum thickness of about 1,800 feet. The subdivisions recognized by Burbank (1933a, p. 145-150)—a lower tuff-breccia member, a latite flow member, and an upper tuff member—were used during the present mapping, except where the Burns quartz latite is greatly altered near the ring-fault zone.

The lower tuff-breccia member is made up of fragments of light-green porphyritic latite in a fine-grained tuffaceous matrix of similar composition. The fragments are generally half an inch to a few inches in diameter and subangular, but some are rounded cobbles and others are angular blocks several feet in diameter in a matrix of smaller angular fragments and sand. Sparsely scattered white to light-pink orthoclase phenocrysts as much as half an inch in length in the fragments help to distinguish these lower breccias from those within the Eureka rhyolite. Burbank found a breccia in Arrastra Basin that is similar to the breccias of the lower member and which is intertongued with the medial flow member. He differentiated this unit, calling it the breccia of Round Mountain, which is included within the lower tuff-breccia member of this report.

The main body of tuff breccia is overlain by interbedded massive flows of very dusky red-purple, dark-gray, and very dusky green latite and andesite, by light-gray and greenish-gray fluidal facies of the same apparent composition but with platy structure, and by breccia beds identical with those lower in the section. This series of massive and platy flows, which forms the middle member of the Burns, reaches a maximum thickness of about 1,000 feet in Arrastra and Little Giant Basins.

Under the microscope, even the freshest appearing rock from massive flows is generally seen to be so altered that identification of many of the constituent minerals is uncertain. All flows examined in thin section were porphyritic with plagioclase phenocrysts 1 to 5 mm long in a finely crystalline matrix. The groundmass commonly shows fluidal texture. The plagioclase is generally albite-oligoclase, but in one flow at the top of Spencer Peak, which may belong to the pyroxene-quartz latite, but is here tentatively included in the Burns, the feldspar is labradorite. Many of the feldspar phenocrysts and much of the groundmass in most flows of the Burns quartz latite are much altered to calcite and sericite; the biotite, hornblende, and pyroxene are invariably altered to aggregates of chlorite, calcite, and iron oxides. No alkali feldspar was identified as phenocrysts in the flows, although some potassium feldspar may be held in the plagioclase or in the finely crystalline groundmass. Quartz is rare in the flows examined, probably amounting to less than 5 percent, and generally occurs with clay minerals in small cavities. Analyses of Burns quartz latite samples are given in table 2.

These analyses, although few and showing considerable variation among themselves, suggest that the Burns quartz latite has, in general, a composition intermediate between Nockolds' average latite and average rhyodacite.

TABLE 2.—*Analyses, norms, and modes of Burns quartz latite, dike rocks, and similar average igneous rocks*

[Analyses for columns 1–4 and 8 were made in the rapid-rock-analyses laboratory of the U.S. Geological Survey under the supervision of W. W. Brannock. Norms for samples in columns 1–6 and 8 were computed from a modification of the chemical analyses in which the percentages of oxides were recomputed on a basis of 0.6 percent H_2O and 0.0 percent CO_2 to remove some of the effect of alteration so that a comparison could be made more easily between these norms and the average norms of fresh rocks reported by Nockolds]

	1	2	3	4	5	6	7	8	9	10
Analyses										
SiO_2	64.1	58.9	57.1	55.0	58.36	56.36	62.09	58.2	54.02	66.27
Al_2O_3	17.1	16.7	16.3	15.9	16.13	15.46	16.77	16.5	17.22	15.39
Fe_2O_3	2.6	4.3	4.3	4.5	3.85	5.00	3.96	3.7	3.83	2.14
FeO	2.3	2.1	3.0	2.5	3.25	1.85	.99	3.2	3.98	2.23
MgO	1.7	2.6	2.7	3.8	2.91	2.00	1.63	3.0	3.87	1.57
CaO	1.0	2.0	5.5	4.5	3.57	5.28	4.26	3.7	6.76	3.68
Na_2O	4.6	3.4	2.2	3.7	4.51	4.91	3.77	3.2	3.32	4.13
K_2O	4.2	4.8	4.3	3.3	2.41	2.73	3.68	4.6	4.43	3.01
H_2O-	1.8	2.7	2.1	3.0	1.61	1.34	.50	2.0	.78	.68
H_2O+							1.32			
TiO_2	.62	.80	.84	.87	1.49	.83	.73	.86	1.18	.66
P_2O_5	.25	.31	.32	.33			.25	.36	.49	.17
MnO	.13	.12	.10	.12	.66	1.49	.14	.19	.12	.07
ZrO_2							Trace			
CO_2	<.05	1.2	2.2	2.8	1.54	1.92		.64		
S							None			
BaO							.10			
SrO							.05			
Li_2O							Trace			
FeS_2										
Total	100	100	101	100	100.29	99.17	100.24	100		
Norms										
Q	16.4	13.7	13.6	7.4	10.8	6.3	15.54	9.8	0.5	20.8
or	25.6	29.5	26.7	20.6	14.5	16.7	21.68	27.8	26.1	17.8
ab	39.8	29.3	19.4	33.0	39.3	42.4	31.96	27.8	27.8	35.1
an	3.1	8.6	22.5	17.8	17.0	12.2	17.79	16.1	19.2	14.5
C	3.1	3.1						.4		
di							1.08			
hy							3.80			
$CaSiO_3$			1.5	1.5	.5	6.2			4.5	1.3
$MgSiO_3$	4.3	6.8	7.0	10.0	7.5	5.1		7.8	9.7	3.9
$FeSiO_3$	1.3		.7		1.5	.9		1.9	2.4	1.3
mt	3.7	5.1	6.5	6.0	5.8	7.4	1.16	5.6	5.6	3.0
il	1.2	1.5	1.7	1.8	2.9	1.7	1.37	1.7	2.3	1.4
hm		1.0		.5			3.20			
ap	.7	.7	.8	.7			.67	1.0	1.2	.3
Modes										
Quartz				0						
Orthoclase										
Plagioclase		(an 10-30)		22(an 10-30)			15(an52)			
Hornblende				13			5			
Pyroxene							1			
Biotite							Trace			
Magnetite				2						
Chlorite										
Apatite										
Groundmass				63			79			

1. Lower tuff-breccia member of Burns quartz latite (46DV57) from gulch above Mighty Monarch mine, about 37°48'26" N., 107°39'20" W. Lab No. 151420.
2. Middle member of Burns quartz latite (46DV9) from Little Giant Basin, about 37°48'10" N., 107°35'50" W. Lab No. 151415.
3. Pyroxene quartz latite or Burns quartz latite (46DV11) from north side of Shenandoah-Dives vein system at about 100 feet below point where vein system crosses ridge southeast of Little Giant Peak, 37°47'48" N., 107°35'51" W. Lab No. 151416.
4. Middle flow member of Burns quartz latite (46DV12) from flow beneath upper member, 37°47'27" N., 104°35'55" W. Lab No. 151417.
5. Burns quartz latite. "Hornblende-andesite, tunnel of Dives claim, upper level, King Solomon Mountain" (Van Horn, 1901, analysis 2), analyzed by E. W. Gebhardt. and W. G. Haldane.

6. Burns quartz latite. "Hornblende-andesite, country rocks of Dives claim, upper level of King Solomon Mountain" (Van Horn, 1901, analysis 6), analyzed by E. W. Gebhardt and W. G. Haldane.
7. Burns quartz latite (P.R.C. 1253) from ridge north of the head of Pole Creek Larson and Cross, 1956, p. 78), analyzed by W. F. Hillebrand, 1905. An erroneous value of 3.18 percent K_2O given by Larsen and Cross has been corrected to agree with the original value reported by Cross, Howe, and Ransome (1905, p. 12). The norm given by Larsen and Cross was based on the corrected value.
8. Intrusive andesite or latite (46DV4) from wide dike in upper part of Swansea Gulch, 37°47'51" N., 107°38'08" W. Lab No. 151513.
9. Latite, average of 42 analyses (Nockolds, 1954).
10. Rhyodacite and rhyodacite-obsidian, average of 115 analyses (Nockolds, 1954)

PYROXENE-QUARTZ LATITE

At the top of Little Giant Peak and along the crest of King Solomon Mountain the upper tuff member of the Burns is overlain by about 200 feet of dense dark-gray and very dusky red-purple flows. These flows are the youngest extrusive rocks preserved in the area. They were originally included within the Burns latite in the Silverton folio (Cross, Howe, and Ransome, 1905); they later were assigned to the pyroxene andesite by Burbank (1933a, p. 151) and are included in the pyroxene-quartz latite of Larsen and Cross (1956). The flows, both in hand specimen and under the microscope, are similar to some parts of the middle flow members of the Burns; but their position, above the tuff that marks the top of the Burns here and in other parts of the quadrangle, makes their correlation with the pyroxene-quartz latite reasonable.

The pyroxene-quartz latite is best exposed in the central part of the Silverton quadrangle, north of the area show on plate 1. The base of the pyroxene andesite, as shown in the Silverton folio on the north side of the Animas Valley above the Shenandoah-Dives (Mayflower) mill, was compared with the pyroxene-quartz latite exposed on King Solomon Mountain. North of the Animas River the sequence of light-colored tuff and platy flows overlain by massive dark cliff-forming flows is sufficiently similar to tentatively confirm correlation with the succession of volcanic rocks on Little Giant Peak and King Solomon Mountain.

As is generally true for all the andesitic rocks of the Silverton volcanic series, microscopic examination of the pyroxene-quartz latite produced little data to supplement field identification. The pyroxene-quartz latite flows on Little Giant Peak and King Solomon Mountain are completely or nearly completely crystalline. Some flows are porphyritic, but in most flows the feldspar crystals range in size from 4 mm or so to submicroscopic crystals. From flow to flow the predominant feldspar ranges in composition from albite-oligoclase to labradorite. No potassium feldspar was noted and quartz is present only in amounts of less than 5 percent. The feldspar is commonly altered to sericite and calcite, and all ferromagnesian minerals are altered to aggregates of calcite, chlorite, and iron oxides. Relict outlines of pyroxene are no more noticeable in this unit than in the Burns quartz latite.

Analyses of pyroxene-quartz andesite from King Solomon Mountain and elsewhere are given in table 3. The norms generally fall within the range between the average norms for Nockolds' doreite and dacite. Sample 46DV8 (column 1) shows, however, a conspicuously high value for or and a low value for an and thus closer affinity with Nockolds' latite (table 2, column 9), although the value for Q is much higher.

TERTIARY INTRUSIVE ROCKS

During and following the extrusion of the volcanic rocks, a more or less elliptical part of the earth's crust north of Silverton subsided 1,000 to 2,000 feet relative to the surrounding area. This block is nearly completely bounded by faults and is about 8 by 10 miles in area. The town of Silverton lies in a valley eroded along the broad fault zone that marks the southern periphery of the subsided block. Not only was the border of the sinking block broken and crushed but also the surrounding rocks were fractured for a distance of several miles away from the peripheral, or ring fault, zone. Although the subsided block is described in more detail in the section on "Tertiary structures," it is mentioned briefly here, because the fissures associated with its formation and development, both those in the peripheral fault zone and the several groups of fractures extending into the surrounding area, were intruded and partly filled by the Tertiary intrusive rocks.

The intrusive rocks are of four general lithologic types: Quartz monzonite, granite porphyry, rhyolite, and andesite or latite. Of these, the quartz monzonite, in the form of stocks, small irregular bodies, and a

TABLE 3.—*Analyses, norms, and modes of pyroxene-quartz latite and similar average igneous rocks*

[Analyses for column 1 were made in the rapid-rock-analyses laboratory, U.S. Geological Survey, under the supervision of W. W. Brannock. Norms for samples in columns 1–3 were computed from a modification of the chemical analyses in which the percentages of oxides were recomputed on a basis of 0.6 percent H_2O and 0.0 percent CO_2 to remove some of the effect of alteration so that a comparison could be made more easily between these norms and the average norms of fresh rocks reported by Nockolds]

Analyses

	1	2	3	4	5	6	7
SiO_2	56.4	55.77	55.68	58.88	56.03	56.00	63.58
Al_2O_3	16.9	16.38	17.09	16.10	15.97	16.81	16.67
Fe_2O_3	4.2	4.27	5.26	3.12	4.78	3.74	2.24
FeO	2.8	4.17	1.98	2.94	3.00	4.36	3.00
MgO	3.1	3.10	2.23	2.30	3.36	3.39	2.12
CaO	3.4	6.66	5.05	6.05	6.44	6.87	5.53
Na_2O	3.8	2.85	2.83	3.17	2.85	3.56	3.98
K_2O	3.9	2.37	1.95	1.86	3.29	2.60	1.40
H_2O-	} 2.8	2.36	1.50	{ 1.66	1.31		
H_2O+				2.86	1.08	.92	.56
TiO_2	.90	.34	1.45	.73	1.01	1.29	.64
P_2O_5	.35			.17	.48	.33	.17
MnO	.16		1.56	.16	.16	.13	.11
ZrO_2				.02	Trace		
CO_2	1.6	1.58	2.97				
S					.08		
BaO				.13	.08		
SrO				.14	.04		
Li_2O				Trace			
FeS_2				.07			
Total	100	100.65	99.55	99.97	99.88		

Norms

	1	2	3	4	5	6	7
Q	9.1	7.7	20.2	17.46	10.86	7.2	19.6
or	23.9	14.5	12.2	11.12	19.46	15.6	8.3
ab	33.0	32.0	24.6	27.25	24.10	29.9	34.1
an	14.7	22.0	26.1	23.35	21.13	22.2	23.3
C	1.3		1.1				
di				3.77	6.51		
hy				5.98	5.50		
$CaSiO_3$		5.1				4.1	1.3
$MgSiO_3$	8.0	8.0	5.8			8.5	5.3
$FeSiO_3$	.3	3.7				3.0	2.8
mt	6.5	6.5	7.7	4.41	6.96	5.3	3.3
il	1.8	.6	2.9	1.37	1.98	2.4	1.2
hm			.2				
ap	1.1			.67	1.39	.8	.3

Modes

	1	2	3	4	5	6	7
Quartz	<5						
Orthoclase	Scarce						
Plagioclase	(Oligoclase)			27 (an58)	34 (an52)		
Hornblende				.5			
Pyroxene				8	9.5		
Biotite				.5			
Magnetite				1	1		
Chlorite							
Apatite							
Groundmass				63	59		

1. Pyroxene-quartz latite, from massive dark flow at south end of the crest of King Solomon Mountain (46DV8), 37°48'03" N., 107°35'34" W. Lab. No. 151414.
2. Pyroxene-quartz latite(?). "Hornblende andesite, Little Giant peak of King Solomon, slightly below summit toward Cunningham Gulch" (Van Horn, 1901, analysis 7), analyzed by E. B. Willard.
3. Pyroxene-quartz latite. "Hornblende-andesite, summit of Little Giant peak of King Solomon Mountain" (Van Horn, 1901, analysis 8), analyzed by E.W. Gebhardt and W. G. Haldane.
4. Pyroxene-quartz latite (P.R.C. 1355) from Vitrophyric pyroxene-quartz latite (P.R.C. 1355) from ridge west of Edith Mountain, Silverton quadrangle, (Larsen and Cross, 1956, p. 80), analyzed by W. F. Hillebrand, 1905. Pyroxene in mode consists of 5 percent augite and 3 percent hypersthene.
5. Pyroxene-quartz latite from Silverton quadrangle, Copper Gulch, on Dolly Varden claim (Larsen and Cross, 1956, p. 80), analyzed by W. F. Hillebrand, 1905. Pyroxene in mode consists of 4.5 percent augite and 5 percent hypersthene.
6. Doreite, average of 38 analyses (Nockolds, 1954).
7. Dacite and dacite-obsidian, average of 50 analyses (Nockolds, 1954).

few dikes, has by far the greatest volume. The andesite or latite, granite porphyry, and rhyolite occur as dikes. The many andesite and latite dikes have an aggregate length of 17 or 18 miles within the mapped area; the outcrops of the granite porphyry dikes and the single rhyolite dike are about 5 miles and a quarter of a mile long, respectively. All these intrusive rocks are younger than the bedded volcanic rocks, but their relative age cannot be determined with certainty, for they are nowhere in contact with one another.

Chemical analyses, norms, and modes of some of the Tertiary intrusive rocks of the Silverton area, together with average chemical analyses of similar rocks, are presented in table 4.

QUARTZ MONZONITE

Several bodies of quartz monzonite and monzonite were intruded into the greatly broken fault zone along the southern and southeastern border of the Silverton caldera. Their general form is irregular, but they tend to be elongated parallel to the fault zone. The largest of these bodies within the mapped area is at the bend of the Animas River just south of Silverton, and it is doubtless continuous with a smaller exposure on the slopes north of Silverton. These two exposures are the east end of the large stock at Sultan and Bear Mountains west of the Animas River, which is described in the Silverton folio (Cross, Howe, and Ransome, 1905). A smaller stock or very wide dike of monzonite crosses Cunningham Gulch about 1 mile south of Howardsville and passes through more than 1,000 feet of volcanic rock on the northeast flank of King Solomon Mountain. The rocks in this area are so thoroughly altered and stained with iron oxides that it is difficult to determine the contact between fine-grained quartz monzonite or monzonite and the rather equigranular facies of the Burns quartz latite. The geologic map shows several much smaller bodies of quartz monzonite along the fault zone, and no doubt others are present beneath the glacial deposits along the sides of the valley; still others may have been not noticed owing to the extensive alteration of the rocks within the ring-fault zone. The following lithologic description given by Cross and Howe (Cross, Howe, and Ransome, 1905, p. 12) for the stock at Sultan and Bear Mountains applies as well to the quartz monzonite exposed on the east side of the Animas River.

The average rock of this great stock is pinkish in color, granular in texture, and of medium fine grain. The eye can readily distinguish in it orthoclase, the pink constituent which tinges the whole mass, white plagioclase, quartz, and dark silicates which are very subordinate in amount. The microscope shows that pale-green augite is the commoner dark constituent, and that biotite and green hornblende are variable

elements. There are the usual number of accessories, including magnetite, titanite, and apatite.

While the feldspars—orthoclase and plagioclase—are often nearly equal in amount and quartz is abundant, so that the rock is to be called quartz-monzonite, there are variations from this composition in several directions. Orthoclase sometimes strongly predominates over plagioclase and the rock becomes a granite, while a corresponding dominance of plagioclase was not observed.

The ferromagnesian minerals increase in amount locally, generally with accompanying fine grain, and a much darker and nearly aphanitic rock results. Oftentimes these darker

TABLE 4.—*Analyses, norms, and modes of quartz monzonite, granite porphyry and similar average igneous rocks*

[Analyses for columns 1 and 5 were made in the rapid-rock-analyses laboratory, U.S. Geological Survey, under the supervision of W. W. Brannock. Norms for samples in column 5 were computed from a modification of the chemical analyses in which the percentages of oxides were recomputed on a basis of 0.6 percent H_2O and 0.0 percent CO_2 to remove some of the effect of alteration so that a comparison could be made more easily between these norms and the average norms of fresh rocks reported by Nockolds]

	1	2	3	4	5	6
Analyses						
SiO_2		63.91	65.88	65.50	73.1	73.66
Al_2O_3		17.07	15.07	15.65	14.2	13.45
Fe_2O_3		4.39	1.74	1.63	.8	1.25
FeO		1.51	2.73	2.79	.88	.75
MgO		.81	1.38	1.86	.63	.32
CaO		4.47	3.36	4.10	.20	1.13
Na_2O		3.48	3.53	3.84	2.2	2.99
K_2O		3.74	4.64	3.01	5.7	5.35
H_2O-		.33			} 2.1	{
H_2O+			.52	.69		.78
TiO_2			.81	.61	.34	.22
P_2O_5		.21	.26	.23	.09	.07
MnO			.08	.09	.04	.03
ZrO_2						
CO_2					<.05	
S						
BaO						
SrO						
Li_2O						
FeS_2						
Total		99.92			100	
Norms						
Q		19.74	18.8	20.0	38.0	33.2
or		21.68	27.2	17.8	34.5	31.7
ab		29.34	29.9	32.5	18.3	25.1
an		20.29	11.7	16.4	.3	5.0
C					4.4	.9
di						
hy		2.00				
CaSiO$_3$			1.4	.9		
MgSiO$_3$			3.4	4.6	1.6	.8
FeSiO$_3$			2.4	2.9	.5	
mt		4.87	2.6	2.3	1.2	1.9
il			1.5	1.2	.6	.5
hm		1.12				
ap		.67	.6	.6	.3	.2
Modes						
Quartz		41.8			13	
Orthoclase		29.4			6	
Plagioclase		14.7 (an₁₃).			10(an₉)	
Hornblende		7.5				
Pyroxene		Trace				
Biotite		2.2				
Magnetite		2.5				
Chlorite		1.8				
Apatite		.1				
Groundmass					71	

1. Quartz monzonite (52–VS–1) from east base of Sultan Mountain.
2. Quartz monzonite (granodiorite) from east base of Sultan Mountain (Larsen and Cross, 1956, p. 228, 232), analyzed by L. G. Eakins, 1905.
3. Hornblende-biotite adamellite, average of 41 analyses (Nockolds, 1954).
4. Hornblende-biotite granodiorite, average of 65 analyses (Nockolds, 1954).
5. Granite porphyry dike (52–VS–5) from northeast slope of Galena Mountain about 37°50′20″ N., 107°33′44″ W. Lab. No. 151521.
6. Calc-alkali rhyolite and rhyolite-obsidian, average of 22 analyses (Nockolds, 1954).

TABLE 5.—*Results of physical tests performed on sample HV-12-46 of quartz monzonite*

[Laboratory No. 73045, E. F. Kelley, Chief, Division of Tests, U.S. Bureau of Public Roads, analyst. Reported July 28, 1947. Description of methods of testing given by Woolf (1953)]

Physical tests

Wear, Los Angles (grading A)_____percent__	18. 9
Toughness_____	14
Absorption_____percent__	. 7
Loss in accelerated soundness test by sodium sulfate solution [1]_____percent__	. 9
Bulk specific gravity (dry basis)_____	2. 73
Compressive strength (1 in. × 1 in. cylinders)__psi__	42, 000

Static immersion stripping test

[24 hours at 140°F]

Coating material	Result
RC-2_____	Unsatisfactory
SC-2_____	Do.
250/300 Pen. AC_____	Do.

[1] A.S.T.M. method C-88-46T. Loss weighted to a uniform grading of material from 1½ inches to No. 4 size.

rocks are near the borders of the mass, but they are not always present in the contact zone. There is sometimes a gradation from the darker monzonite into the common facies, and in other places dike contacts were seen.

Complementary to the facies richer in darker silicates are small dikes of aplitic granite or quartz-monzonite, almost free from augite or other ferromagnesian minerals. These dikes are variable in grain from very fine to coarse, and an irregular porphyritic texture is also often developed.

The texture of the part of the stock exposed south of Silverton is uniform and the average size of the grains is about 2 mm. It is generally light to medium gray or pinkish gray and mottled, owing to clusters of hornblende grains and iron oxide a few millimeters across. Although the description quoted gives augite as the commoner dark constituent, only a trace was found in microscopic examination of a specimen collected from the stock at the east base of Sultan Mountain. Analyses of this rock are given in table 4.

The quartz monzonite is a sound, strong, durable rock. The results of tests performed by the U.S. Bureau of Public Roads on a sample collected by Helen D. Varnes 1.2 miles south of Silverton on U.S. Route 550 are shown in table 5.

ANDESITE AND LATITE

Andesite and latite dikes occupy several well-defined groups of fractures within an east-west belt 2 or 2½ miles wide south of the ring-fault zone. Although the dikes vary somewhat in texture and in proportions of feldspar, hornblende, augite, and biotite, they all appear to be latite or andesite. The groundmass is predominantly light green to dark grayish green grading to shades of reddish purple. Most of the phenocrysts are gray plagioclase and are a few millimeters long. Hornblende occurs as single small crystals or in clusters and is generally altered to chlorite. Products of late magmatic or hydrothermal alteration—such as chlorite, calcite, sericite, iron oxides, epidote, and clay-

like minerals—are widespread. The various dikes or parts of dikes differ somewhat in appearance, but this may be due more to differences in degree of alteration than to differences in original composition. Many of the dikes are so similar in appearance to the flows within the Burns quartz latite that they are often difficult to trace through the flows unless the dikes are fractured or veined. The dikes are commonly 20 to 40 feet thick.

Burbank identified two groups of dikes within Arrastra Basin (1933a, p. 152–154); one group strikes about N. 80° E. and the other, between N. 45° W. and N. 70° W. The present mapping shows that some of these dikes extend considerably beyond the area that he mapped and that a third group of fractures and latite dikes with a general trend N. 45° E. is in the area to the west of Arrastra Basin.

The group that strikes N. 80° E. includes two prominent nearly parallel dikes, the Arabian Boy and the Arrastra. These dikes are well exposed in Arrastra Basin, where they are about 1,200 feet apart. The northern dike, the Arabian Boy, can be traced for almost 4 miles, from Cunningham Gulch westward, with some discontinuities, nearly to the trail south from Silverton to Kendall Gulch. Burbank (1933a, p. 153) describes the Arabian Boy dike as feldspathic and somewhat porphyritic, closely resembling some latitic flows of the Burns. It contains hornblende and is locally amygdaloidal. The dike is 30 to 35 feet wide in Arrastra Basin but narrows both eastward and westward to 10 feet or less.

The Arrastra dike is a feldspathic andesite and has been traced for about 8,500 feet, from the ridge northwest of Kendall Peak eastward to talus in Dives Basin. It is 85 feet wide where it extends across the ridge northwest of Kendall Peak but tapers to a few feet near the head of Dives Basin. Both the Arabian Boy and the Arrastra dikes dip 75–85° N. in the vicinity of Arrastra Basin, but the Arabian Boy dike appears to become less steep near Cunningham Gulch.

The dikes, with trends of N. 45°–75° W., mapped by Burbank are the Mayflower dike, followed by part of the Shenandoah-Dives vein system, the Magnolia dike, the Silver Lake dike, and the dike followed by the Titusville vein and its eastward extension, the Buckeye vein.

A chemical analysis of one specimen of andesitic or latitic dike rock is given in table 2.

GRANITE PORPHYRY

A group of granite porphyry dikes extend across Cunningham Gulch near the Highland Mary mill, continue southwestward into the western part of Spencer Basin and thence westward along the ridge north cf Deer Park Creek, where the outcrops end. A dike also is exposed on the east side of Cunningham Gulch, east of the area covered by this report; it crops out on the northeast flank of Galena Mountain, a few miles farther to the north and northeast, near the ring-fault zone. Altogether the dike or dikes are exposed intermittently over an arcuate course of about 7 miles.

Only one dike is exposed on the east side of Cunningham Gulch and also at the southwest extremity on the ridge north of Deer Park Creek. There are several overlapping or en echelon dikes in Spencer Basin, and at two places on the ridge north of Spencer Basin the dikes swell to masses several hundred feet wide. A dike dips steeply eastward on the northeast flank of Galena Mountain; through the remainder of its course it is generally vertical or dips steeply toward the concave side of the arc.

The granite porphyry is light greenish gray and is described by Cross, Howe and Ransome (1905, p. 11) as follows:

This rock is a typical felsophyre, having abundant phenocrysts of orthoclase and quartz, with a much smaller number of biotite and plagioclase, in a felsitic groundmass of orthoclase and quartz. This groundmass is usually very finely and evenly granular, the two minerals being sometimes interlocked in graphic intergrowth.

The dike is 20 to 30 feet wide in most places and has very marked porphyritic texture in the center, with contact zones of dense, greenish felsite, carrying a few quartz phenocrysts, the zones being 3 or 4 feet wide. The green color is due to minute particles of chlorite which may be an infiltration product from the adjacent volcanic rocks.

Examination of the granite porphyry in thin section yielded the additional information that the composition of the plagioclase is about An_9, that both orthoclase and sanidine are present, and that the small prismatic spots of chlorite are apparently due to alteration of hornblende. The rock resembles none other in the area and, owing to its light color, may be easily seen where it cuts the darker schist and volcanic rocks. An analysis of this rock is given in table 4.

FIGURE 7.—View southeastward from Cement Creek across the Animas Valley toward Swansea Gulch. Swansea Gulch lies between the two prominent peaks, Kendall No. 2 on the right and Kendall No. 3 on the left. The upper cutting limit of the last alpine glacier in Swansea Gulch is marked by the top of the cliffs on the east side of the high basin. The moraine-mantled slopes on the south side of the Animas Valley are largely timbered. The sharp V-shaped ravine in the lower part of Swansea Gulch is cut in the moraine and underlying rock near the center of the photograph, and gives some measure of postglacial erosion.

RHYOLITE

A rhyolite dike, trending N. 15° W., crops out discontinuously for about 1,300 feet on the west side of Cunningham Gulch, just south of Howardsville. The dike is vertical, about 20 feet thick, and white except where stained by weathering of disseminated pyrite. It is nonporphyritic and very fine grained. Flow lines dip steeply to the north. Its relation in age to the other intrusive rocks is unknown.

QUATERNARY DEPOSITS

The surficial materials of Pleistocene and Recent age within the area have been divided into four groups for the purpose of mapping: glacial till, alluvium, talus, and landslide material.

The most extensive of the Quaternary deposits is glacial till produced largely during the last stage of glaciation. According to Atwood and Mather (1932) the glaciers of Wisconsin time covered the area to within a few hundred feet of the tops of the peaks and higher ridges. The upper cutting limit of the ice is well shown in figures 3 and 7 by the top of the abrupt cliffs along the sides of Swansea, Blair, and Arrastra Gulches. The slopes above the cliffs rise more gently to the crests of the peaks. Moreover, the deep oxidation of many of the veins on these high slopes indicates lack of glacial scour, at least during the last advance of the ice. The depositional products of this glacial erosion do not form conspicuous topographic features in the smaller basins and gulches. Much of the material was either transported out of the area by ice or reworked and removed by streams. Large blanketlike deposits of till were, however, left by the ice along the sides of Animas Valley and at the mouths of the large gulches. Figure 2 shows the low flat-topped bench of a lateral moraine just to the left of the alluvial flat on which Silverton is built. On the far side of Cement Creek a road extends through a hummocky deposit of moraine. The moraine on the south side of the Animal Valley is covered by a dense growth of timber.

Much of the glacial material has been reworked by present streams and deposited, together with the products of current erosion, as fans or stream alluvium in the larger valleys. The fans and valley bottoms are shown as alluvium on the map, but the distinction between alluvium and the lower parts of the moraine could not be sharply drawn except where both were clearly exposed. For the most part, the areas shown as talus consist of bare sloping accumulations of loose angular rock fragments and interstitial fine material beneath cliffs. Talus was mapped where it is so thick that it might hamper mining operations. Many tongue-shaped rock glaciers in the high basins were also mapped as talus. They are unusually thick, perhaps locally more than 100 feet, and generally contain considerable interstitial ice not far beneath the surface. The problems of maintaining adits through these masses, owing to the constant slow movements, are well known to miners of the San Juan region.

Since retreat of the last glaciers, several of the larger streams have in some places eroded through the moraines and glaciofluvial deposits and cut narrow gorges, ranging from 75 to 200 feet in depth. Two such gorges are the short inner canyon of the Animas near the Shenandoah-Dives mill, shown in figure 8, and the V-shaped ravine along the lower part of Swansea Gulch, shown in figure 7.

STRUCTURE

The brief outline of geologic history indicates that the structural record of the western San Juan Mountains is long and complex. Mountain building has affected this area intermittently since very early in geologic time, and the structural pattern of each period of deformation has controlled in some degree the pattern of the next. In turn, the patterns of early deformations are obscured by those of succeeding periods or are hidden by subsequent deposits, so that it generally becomes more difficult to decipher the structural features as progressively older rocks are studied.

PRECAMBRIAN STRUCTURES

However obscure the earliest geologic history of this region may be, studies near Ouray (Burbank, 1940; Cross, Howe, and Irving, 1907), and especially to the south of Silverton (Cross, Howe, Irving, and Emmons, 1905), have shown clearly that the folding, faulting, and metamorphism produced by several periods of deformation in Precambrian time in this area were of an intensity not approached by any later epoch of mountain building.

Banding and foliation are the dominant structural features of the Precambrian schist and gneiss exposed near Silverton. The large bands, a few feet to many tens of feet wide, probably result from variations in composition of the original rocks. Smaller bands and foliae were produced by recrystallization of the original rocks under high heat and pressure. This foliation and schistosity was produced by Precambrian deformation, for in many places Cambrian quartzite lies undisturbed upon the older contorted schist and gneiss.

The strike of banding and foliation within the Precambrian rocks, where they are exposed in the southern part of the area, arcs from about N. 80° W. near

FIGURE 8.—The shallow but precipitous inner canyon of the Animas River viewed westward toward Silverton from a point near the Shenandoah-Dives tram and mill. Moraine is exposed among the trees on the left. Shattered rocks typical of the ring-fault zone are visible on the right and in the middle distance.

the Animas River to west in the headwaters of Deer Park Creek and in Spencer Basin and to northeast along the west side of Cunningham Gulch. The dip is generally steep and to the north but is to the south in some places and is complicated by folding, especially in the Animas canyon. The most prominent joints dip steeply and strike at right angles to the foliation. A few are slightly mineralized, such as those on the north slope of Sugarloaf Mountain and at the headwaters of Deer Park Creek.

PALEOZOIC AND MESOZOIC STRUCTURES

Paleozoic and Mesozoic strata are lacking in the higher parts of the western San Juan Mountains, owing to broad doming of the entire region at the close of Cretaceous time and subsequent erosion to a surface of low relief. The Telluride conglomerate of Oligocene(?) age was deposited upon this surface and thus rests on progressively older rocks as the Precambrian core of the dome is approached. The present San Juan Mountains include and extend somewhat beyond the area of this uplift. Most of the remaining pre-Tertiary strata are in a belt around the outer flanks of the range, but enough of the sedimentary section is preserved and exposed in the deeper canyons within the mountains to indicate some of the early structural history of the region.

Oscillations in sea level are recorded in the rocks through much of the Paleozoic era. Periods of no deposition or of uplift and erosion in this part of Colorado are indicated by the absence of strata of Early and Middle Cambrian and of Ordovician through Middle Devonian ages. Uplift and deep weathering at the close of the Mississippian produced an irregular pitted and cavernous surface on the Leadville limestone, which was covered by the Molas shale of Pennsylvanian age. The Paleozoic uplifts apparently produced broad areas of low relief and were not accompanied by recognizable faulting and folding.

Upper Paleozoic and Mesozoic strata are not present in the area covered by plate 1 but crop out a few miles to the west in a continuous sequence without angular unconformity. All the Paleozoic and Mesozoic strata dip to the west; and on the higher ridges west of the Animas Canyon strata of Permian to Cretaceous age are overlapped by Telluride conglomerate. The overlap of the Telluride conglomerate into westward-dipping Paleozoic strata can be seen at the right edge of figure 4.

Although the general westward dip of the Paleozoic section was probably produced by the uplift in Late Cretaceous time, it is possible that earlier uplifts caused some of this dip. At Ouray, Burbank (1940, p. 197) found an angular unconformity between Triassic and Permian strata and also axes of pre-Triassic folds and faults; therefore, late Paleozoic doming affected at least that part of the western San Juan Mountains. The principal flexures and faults related to late Paleozoic movements are much closer to the Precambrian core of the mountains than is the main monocline produced in the Late Cretaceous. Late Paleozoic doming might have affected the area southwest of Silverton but evidence of unconformity between Triassic and Permian strata would have been removed by erosion of the Telluride conglomerate prior to deposition of the San Juan tuff or by erosion of strata during cutting of the Animas Canyon.

Doming at the end of the Cretaceous period may have been accompanied by some faulting. The faults along which a block of Ouray and Leadville limestones dropped into Precambrian rocks just south of the Highland Mary mine are older than San Juan tuff, and may have been active in this Late Cretaceous uplift.

TERTIARY STRUCTURES

SILVERTON CALDERA AND SUBSIDED BLOCK

Structural features that developed during Tertiary time had profound control upon the distribution and localization of later ore deposition within the San Juan "triangle" formed by the towns of Ouray, Telluride, and Silverton. The most important feature is the Silverton caldera to which most of the ore deposits of this region bear close relation.

Although some of the faults that bound the caldera had been shown in the early folios, the caldera structure and the subsided block remained unrecognized until Burbank began detailed study of the adjacent mining districts. He found that the structural elements of late Tertiary centers of mineralization were closely related to the central area of subsidence. The concept was first developed in his report on Arrastra Basin (Burbank, 1933a), and was supported by further

mapping at Ouray (Burbank, 1940), in the area to the northwest of the area of subsidence (Burbank, 1941), to the north (Kelley, 1946), to the south (Varnes, 1948), and on the northeast side (Burbank, 1951).

The Silverton caldera is the ancient depression occupied by the Silverton volcanic series. All the volcanic-rock units occupying the caldera are much thicker inside its rim than outside. The area of accumulation of the Silverton series is roughly elliptical in outline and about 15 by 30 miles, including both the Lake City and Silverton volcanic centers. Within the west-central part of the caldera a block about 8 miles by 10 miles, also of elliptical outline, has been faulted down 1,000 to 2,500 feet. As shown on figure 9, this subsided block is bounded by a belt of closely spaced, steeply dipping normal faults—the ring-fault zone. Radial fractures also formed outward from the rim of the subsided block, principally in zones directed toward intrusive bodies a few miles away, as at Stony Mountain and Ophir Valley, and toward another area of subsidence to the northeast near Lake City.

The subsided block in the Silverton caldera is not reflected by the present topography, for the basin apparently was filled almost continuously with volcanic material as it formed. Since the period of volcanism, the subsided block has been subjected to several cycles of erosion by water and ice so that its outline on the west, south, and east is marked by deep curving valleys excavated into the peripheral, or ring-fault, zone. The ring-fault zone is followed on the west by Mineral Creek southward from Red Mountain Pass to Silverton and on the east and south by the Animas River from Eureka to Silverton.

In the San Juan Mountains, there is now an enormous volume of volcanic material at or close to the earth's surface which once lay at great depths. To take the place of the matter extruded, it must be assumed that the earth's surface gradually subsided. No large void could exist, and probably no extensive subcrustal region could remain long under even moderately reduced pressure without rupture of its roof. Rock material appears to have migrated under the influence of unknown forces from beneath a fairly large area lying to the northeast of Silverton. It may have forced its way upward and extruded as lava, or moved laterally and eventually upward to form the many small stocks of igneous rock a few to a dozen miles away. The overlying rocks then bowed down and finally failed under their own weight.

The ring faults that bound the subsided block are either vertical or dip steeply inward, so far as can be seen. The block may thus have acted like a slightly

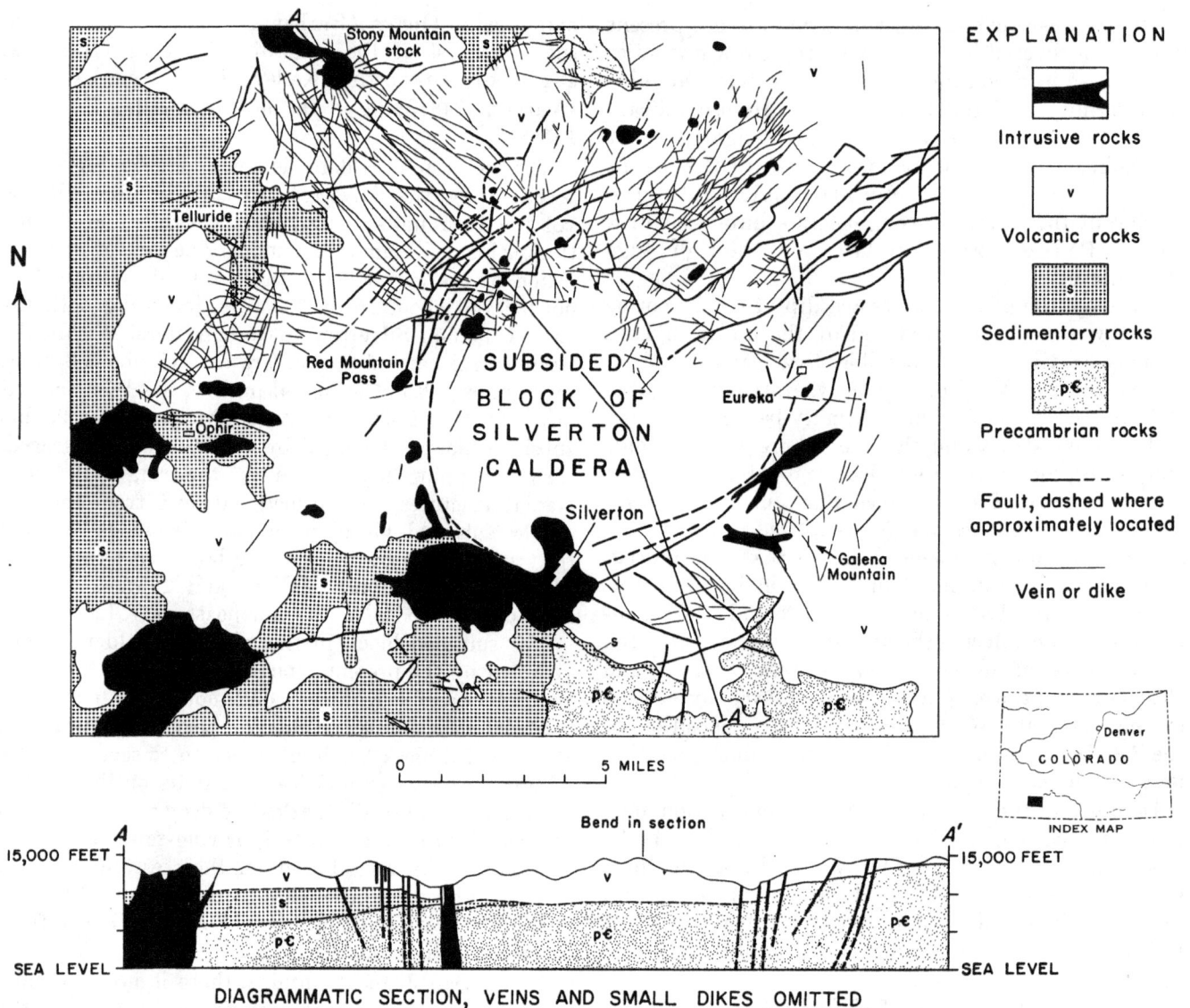

FIGURE 9.—Generalized geologic map of the part of the Silverton caldera area showing the subsided block. Adapted from Burbank, Eckel, and Varnes (1947, pl. 28).

tapered plug as it sank and subjected its periphery to strong horizontal radial compression. Actually, the downward movement of the block was relatively small, probably not more than one-twentieth of its diameter, but it was enough to cause permanent deformation of the rocks as much as several miles from the border. The downward movement and wedging was probably not an even continuous process, but took place in several or many stages. The release of pressure at the base of the block may have had some cyclic rhythm, or the block may have "hung up" for a time, owing to friction along the ring faults until shear stresses due to gravity exceeded shear resistance. Some lateral pressure may also have been exerted by small intru-

sions into the block itself. Lateral pressure on the periphery of the subsided block was probably periodically augmented by wedging action of a graben at the northeast corner.

Evidence in Ophir Valley, to the west, indicates that regional downwarping began between the time of deposition of the Telluride conglomerate and the first accumulation of volcanic material. The conglomerate grades into shale to the west and was therefore presumably deposited upon a westward-sloping surface. But at Ophir Valley the conglomerate dips eastward toward the caldera and is overlain by the nearly horizontal San Juan tuff and flows of the Silverton series. These volcanic rocks thicken abruptly to the east. Subsidence continued, accompanied perhaps by minor faulting, throughout the period of volcanic eruptions. Large-scale faulting around the rim of the area of subsidence took place, and many intrusive bodies of monzonite, diorite, and related rocks were injected into the ring-fault zone and the volcanic flows in the periphery. Continued movements, related to sinking of the block and injection of the surrounding stocks, produced new fractures, some of which were filled with dikes. Many of these fractures also formed channels for later ore deposition.

The dominant structural features in the peripheral area to the south are the ring-fault zone and several systems of shear and diagonal tension fractures.

RING-FAULT ZONE

The ring-fault zone is 1 to 1½ miles wide near Silverton. The total vertical displacement across this zone can be estimated from the difference in altitude of the base of the pyroxene-quartz latite between the north and south sides of the Animas Valley. The base of the pyroxene-quartz latite on the north side is at an altitude of 11,250 feet. On the south side, at the northernmost exposure on King Solomon Mountain, it is at about 13,000 feet. The geologic map in the Silverton folio shows the base of the pyroxene-quartz latite along the north side of Animas Valley to be fairly regular and tilted slightly to the west, although the base may be shown too low by several hundred feet. Assuming that the contact between Burns quartz latite and the pyroxene-quartz latite was originally level, the total displacement across the fault zones in the Animas Valley is about 1,750 feet. This is about 600 feet less than Burbank's estimate (1933a, p. 158).

Concerning the age of the ring faults relative to ore deposition, Burbank states (1933a, p. 158–159):

The greater movement on the faults of the Animas system must have been of premineral age, as all major faults and the fissures paralleling them show alteration of the walls similar to but more intense than that of walls adjoining the nearby veins. Quartz, sericite, pyrite, epidote, chlorite, and specularite in different proportions constitute the principal alteration products along the faults. Some movement of post-mineral age or occurring late in the epoch of mineralization is indicated by the displacement of veins by the faults and by mineralized material that has been sheared and abraded by friction within the faults.

Displacement in the ring-fault zone was distributed along many individual faults. The faults are commonly marked by the zones of strong alteration, sheeting parallel to the faults, and rubble cemented by gouge. Their courses are often shown by a small notch or long narrow swales as they cross the higher ridges. The lower slopes south of the Animas River are covered in large part by talus, moraine, soil, and timber that undoubtedly conceal many more faults than are shown on the map. Rocks within the fault zone are generally so broken and stained that Burns quartz latite has not been subdivided within the main part of the fault zone. The rocks within the fault zone are more easily eroded than the less shattered rock to the south. This condition has caused a break in slope at the south edge of the ring-fault zone along several of the prominent ridges on the south side of the Animas Valley, as shown in figure 10.

Displacement along individual faults was determined by Burbank at a few places on the south edge of the fault zone where either the lower part of the Burns or the Eureka rhyolite were clearly faulted. One fault with an estimated throw of 700 to 800 feet and another with 200 feet of throw are exposed on the east side of Arrastra Gulch. Three closely spaced faults of 75, 125, and 500 feet of throw are exposed on the west side of the gulch.

The major ring faults do not cut the larger monzonite and quartz monzonite intrusions, which are fresh and unaltered in their interiors. The outer parts of the larger intrusive bodies and the whole of the smaller ones are considerably stained and broken as a result of minor movement and alteration along the ring faults during the period of mineralization.

FRACTURING OUTSIDE OF THE RING-FAULT ZONE

GENERAL FEATURES

The production of ores in the area southeast of Silverton has come almost entirely from veins in a complexly fractured zone that lies to the south and southeast of the ring faults. The most intense fracturing and heaviest mineralization are confined to a roughly triangular area, shown in figure 11, which is bounded by the ring-fault zone on the north, by the Titusville vein on the southwest, and by an arcuate shear zone (3a) on the east and southeast. Geologic data in the eastern and western quarters of figure 11 are taken from the Silverton folio (Cross, Howe, and Ransome, 1905); data for the central part are derived from the geologic map that accompanies the present report. The courses of many faults, veins, and dikes have been modified somewhat before plotting on figure 11 to remove the effect of topography in order to better show true strike.

Three systems of fractures may be recognized within the triangular area: (1) a system that is approximately concentric with and 1½ to 2½ miles south of the southern border of the caldera; (2) a system of shear fractures and related tension fractures in the

FIGURE 10.—View southwestward from a point near Howardsville toward King Solomon Mountain. The Pride of the West mill is at the right. Cunningham Creek enters the Animas Valley from the left in the middle distance. Greatly broken rock in the southern border of the ring-fault zone crops out on the west side of Cunningham Gulch below the timber. The break in slope somewhat above timberline on the north slope of King Solomon Mountain marks the southern border of the ring-fault zone.

western half of the triangle; and (3) a system of shear fractures in the eastern part of the triangle.

The second, or western shear system, includes three groups of fractures. Each group can be identified by its trend (see fig. 11) : a northwestward-trending group of shear fractures (2a), a northeastward-trending group of shear fractures (2b), and a diagonal group of tension fractures trending somewhat west of north (2c). The eastern shear system is also tentatively divided into three groups: the eastern arcuate group of shear fractures, partly filled with granite porphyry dikes, which extends from Galena Mountain to Kendall Gulch and includes several other northward-trending fractures lying east of Cunningham Gulch (3a); northwestward-trending fractures on both sides of Cunningham Gulch (3b), and fractures that apparently are approximately radial to the ring-fault zone (3c). There are, in addition to these seven main sets, numerous smaller fractures, narrow veins, and

a few dikes that do not fit into any easily recognized system or group. These apparently are of less significance, both in the regional structure and in formation of the mineral deposits.

CONCENTRIC SYSTEM

The concentric system consists of several major fractures and dike-filled fissures that trend nearly parallel to the ring-fault zone. The most prominent of these are the Arrastra and Arabian Boy andesite dikes (pl. 1), which crop out about 1½ miles south of the ring-fault zone. A narrow but persistent slightly mineralized fracture that crosses Arrastra Creek at an altitude of about 11,700 feet may also belong to this fracture system. There is some question whether the andesite dikes on the ridge south of Kendall Gulch should be included in the concentric system, for these dikes dip steeply to the south, whereas the Arabian Boy and Arrastra dikes dip to the north.

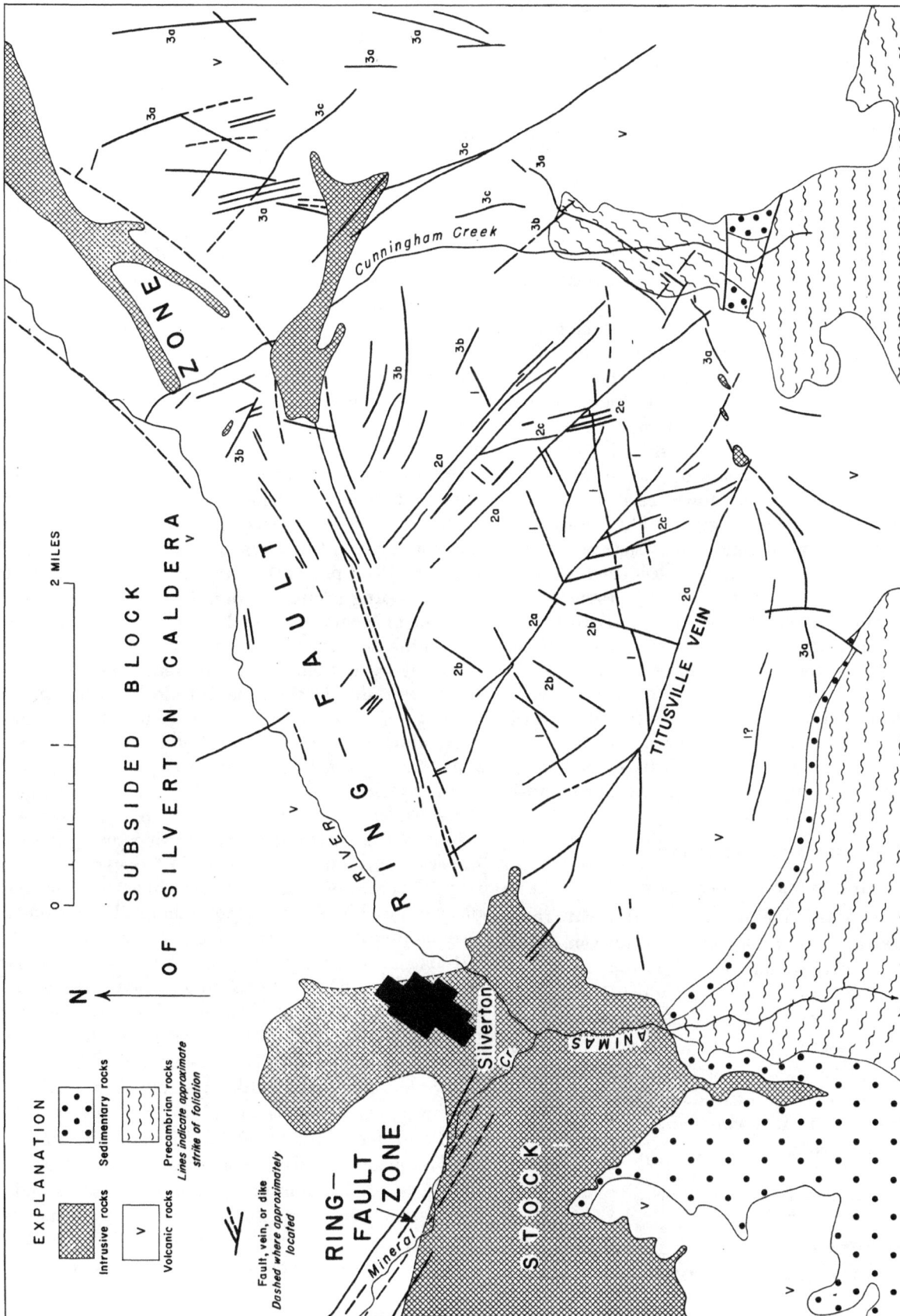

FIGURE 11.—Map showing principal fracture systems of the South Silverton mining area. The courses of the fractures have been somewhat modified from the outcrops, shown in plate 1, to remove the effect of topographic relief. (1) Concentric system; (2) western shear system consisting of (2a) northwestward-trending faults, (2b) northeastward-trending faults, and (2c) diagonal tension fractures; (3) eastern shear system consisting of (3a) eastern arcuate group, (3b) northwestward-trending fractures, and (3c) fractures that are approximately radial to the ring-fault zone.

They also curve somewhat more than the others. These variations, however, may reflect control of fracturing by the structure in the underlying Precambrian schist and gneiss, which here lie at a relatively shallow depth beneath the cover of volcanic rocks.

The age of the concentric fracture system relative to the western shear system cannot be determined by the field relations. Both systems of fractures may have been injected simultaneously with andesite dikes. Movement on the northwestward-trending group certainly continued long after the concentric fractures were filled, but there is little evidence concerning which fractures first started to form.

Dikes and fractures of the concentric system rarely show close jointing or brecciation along their walls. Horizontal or vertical movement in the plane of the fractures cannot be detected, hence the fractures are apparently due to tension exerted at right angles to their present trends. One can only guess how this stress was applied. Perhaps movements of the central area of subsidence, either up or down relative to the periphery, produced bending stresses that were relieved by the formation of concentric fractures in the bordering rocks. Long-continued subsidence of the central block may also have periodically lessened the radical stress imposed on the periphery, if filling in the block did not keep pace with subsidence. This might have allowed the rocks in the periphery periodically to expand slightly inward and fracture in concentric circles. If the subsided block was forced upward slightly by a resurgence of magmatic pressure the radial stress on the periphery would be reduced and dike material could have been injected along lines of least principal normal stress.

WESTERN SHEAR SYSTEM

NORTHWESTWARD-TRENDING GROUP

The northwestward-trending group of shear fractures includes the most prominent dike or vein-filled fractures in the district, such as the Titusville, Silver Lake-Royal Tiger, Mayflower-Shenandoah-Dives, Big Giant, and Black Prince-Gold Lake veins and the Magnolia dike. Fissures of the northwestward-trending group diverge considerably in strike and dip, as may be seen from the average attitudes of the six main fractures given in the following table in order from southwest to northeast:

	Strike	Dip
Titusville vein	N. 71° W.	70°–80° NE.
Silver Lake-Royal Tiger vein	N. 57° W.	65°–80° NE.
Magnolia dike	Curving.	65° NE.
Shenandoah-Dives vein system	N. 43° W.	70°–80° NE.
Big Giant vein	N. 42° W.	70°–80° SW.
Black Prince-Gold Lake vein	N. 45° W.	55°–80° SW.

The inclusion of these fissures into one group, if based only on their general northwestward strike, might well be questioned. However, the movements that have occurred along these several fissures are similar and suggest, more strongly than the approximate concurrence in trend, that the fractures are related in origin. Unlike those of the concentric system, fissures of the northwestward-trending group show evidence of considerable shear movement, both horizontal and vertical. The relative horizontal displacement is in the same sense along each fracture. If an observer faces the fault, the block on the opposite side appears to have moved to the right relative to the block on which he stands. The final vertical movement has been downdip, or "normal," along all the northwestward-trending fractures. Where the horizontal and vertical components of movement can be measured, they generally are comparable and of the order of 100 to 300 feet.

Several classifications of the northwestward-trending fractures and theories as to their origin have been proposed. The divergence in strike of the fissures in the northwest group was noted many years ago by Prosser (1914, p. 1229), who described the northwestward-trending group as "radially disposed around the junction of Mountaineer and Cunningham Creeks," near the southeast corner of the mapped area (pl. 1). Burbank regarded them, in a different sense, as tension fractures radial to the subsided block (1933a, p. 182), although he recognized that the Titusville fracture, in particular, diverges considerably from a radial trend. I previously called the northwestward-trending group a "fan system" opening toward the southern border of the subsided block (Varnes, 1948, p. 8). Because the origin of the northwestward-trending group is of key importance to an understanding of the structural history of the mining district, Burbank's earlier theory of origin and a revised hypothesis resulting from further mapping and analysis are given below.

Burbank suggested that radial fractures associated with the area of subsidence might have resulted from tension due to bulging of the earth's crust by intrusion of igneous magmas (1933a, p. 176). His later map of the area to the northwest of the caldera adds the possibility that some radial-dike systems may result from downwarping rather than bulging of the crust (1940, p. 213). Burbank included the Silver Lake, Magnolia, and Shenandoah-Dives fractures of Arrastra Basin within the radial tension system of fractures and believed that they began to form almost simultaneously with the concentric fractures that trend N. 80° E. He ascribed the horizontal displacement, observed along northwestward-trending fractures that traverse dikes of the concentric system, to later forces. These forces were either differential lateral pressures originating in an area to the northwest, near or within the subsiding central block (1933a, p. 183–184), or compression directly from the north (1933a, fig. 2e and p. 199).

The principal difficulty in assigning the northwestward-trending fractures of the South Silverton area to a radial system is that they depart considerably from a radial trend. If the Titusville, Silver Lake, Shenandoah-Dives, and Black Prince-Gold Lake fractures are projected northwestward along their average strike, they do not intersect the main trend of the ring-faults along the southern border of the area of subsidence at near 90°, as radial fractures should, but at angles of about 38, 51, 65, and 63°, respectively. Moreover, the divergence previously noted, especially between the Titusville fracture and the others, is opposite to the divergence that a system of fractures radial to the caldera would have. There is also no direct evidence that the northwestward-trending group originated by fracturing due to tensile stress, although such evidence would probably be obliterated by any later strong shearing movement.

A "compressed-wedge" hypothesis is here proposed to account for fractures in the western shear system. According to this hypothesis, all northwestward-trending fractures originated not by tension but by shear stress due to pressure from the area of subsidence to the north. The Titusville fracture formed somewhat earlier than the other fractures so that a huge wedge was formed at an early stage in the deformation. The wedge was bounded on the north by the ring-fault zone and on the south by the Titusville fracture. Its apex is directly south of Silverton, near the bend in the Animas River. The wedge forms the western part of the mineralized triangular area shown in figure 11.

The northwestward- and northeastward-trending shear fractures within the wedge were produced as the wedge was compressed. As shearing continued, the previously unbroken strips of rock between the principal faults were torn obliquely by numerous tension fractures of the diagonal group.

The concept of the Titusville fracture forming somewhat earlier than the other northwestward-trending fractures is only an assumption. There is no visible field evidence, such as displaced intersections, to support this assumption, but several lines of indirect evidence indicate that the Titusville is the master fracture of the western shear system. If the Titusville is projected to the northwest, its trend would follow the straight part of the valley of Mineral Creek that extends northwestward from the junction of Mineral Creek and the Animas River. This segment of the valley, in turn, probably follows the fractures that defined the southwest border of the subsided block, although near Silverton the ring faults themselves were obliterated by the Bear Mountain intrusive stock. The Titusville is a major shear fracture, which appears to have grown outward from the highly crushed southwest corner of the subsiding block along a continuation of the ring-fault zone on Mineral Creek.

Because the origin of the Titusville fracture seems to be somewhat more closely connected with the movements of the subsiding block than the other shear fractures, we may infer that it formed a little earlier. It is also the longest of the northwestward-trending fissures. Perhaps the most convincing evidence for early formation of the Titusville is that fracturing south of it is markedly less intense than that to the north. Few veins of economic value have been found within Kendall Gulch, in the drainage of Deer Park Creek, or in Spencer Basin. The Titusville appears to have exerted strong control over fracturing in the district, and to have done so it must have formed at an early stage. In this and in the subsequent discussion of other fractures, the reader should bear in mind that the timing refers only to the initiation of breaks in the earth's crust and not to the sequence of later injections of vein filling into the fractures.

Once the Titusville fracture had been formed, the forces directed outward from the subsiding block no longer acted upon a continuous body of rock around the southern border, unbroken except by a few concentric fractures, but rather on a side of the wedge. As soon as it was formed, the Titusville fracture undoubtedly altered the distribution of stresses around the southern border of the block from a rather simple state of radial compression to one considerably more complex.

The analyses of stresses and theoretical directions of faulting in the wedge are taken up in more detail in chapter B. The pattern is that of pairs of conjugate shears. In the western shear system the two parts of the pair are the northwestward-trending group and the northeastward-trending group.

NORTHEASTWARD-TRENDING GROUP

Northwestward-trending fractures dominate the western shear system, but a few shear fractures of the complementary northeastward-trending group have been found. One of particular interest is the partly concealed fault in Woodchuck Basin shown near the center of plate 1. The horizontal component of displacement along this fault, as indicated by offset of the Arabian Boy and Silver Lake dikes, is left lateral or opposite to the movement along the northwestward-trending group. The fault may continue southwestward into Kendall Gulch, through a deep cleft in the ridge west of Kendall Peak. Talus covers the intersection of this fault and the Titusville vein. The plan map of the Titusville workings suggests, however, that a fault with left-lateral displacement was intersected underground on the 7th level (the level extending farthest west on plate 2) at the point where the drift makes an abrupt jog. The relative horizontal positions of this point underground, the cleft in the ridge, and the course of the fault, as it may be traced northward down into Woodchuck Basin, indicate that the fault dips steeply to the northwest. If this fault and the Silver Lake fracture are regarded as conjugate shears, then the indicated direction of compression is roughly north-south.

Another northeastward-trending shear fracture with left-lateral displacement cuts the Melville vein, as shown by Burbank in a detailed map of the fissures west of Silver Lake (1933a, fig. 2b). The position of this fault on plan maps showing levels of the Melville mine indicates that the fault also dips steeply to the northwest. Several other fractures, which are in part dike filled and which trend across Blair Gulch, are tentatively assigned to the northward-trending group.

DIAGONAL GROUP

The third group within the western shear system consists of many fissures that trend diagonally across the long narrow blocks bounded by the northwestward-trending group of shear fractures. The diagonal fractures generally strike N. 10°–30° W. but swing more northwestward or southeastward as they approach fractures of the main northwestward-trending group.

The most prominent diagonal fractures are those in the footwall of the Silver Lake fissure between Silver Lake and Woodchuck Basin. Fissures here are confined mainly to a few strong zones that include some of the most productive veins of the district, such as the New York, Iowa, Royal, Stelzner, Black Diamond, and Melville veins. Similar fractures diverge into the footwall of the Shenandoah-Dives fissure on Little Giant Peak (see figure 25), but the fractures here are more dispersed and complex and the veins are narrow and unproductive. Several veins in Little Giant Basin—the Mountain Quail, North Star Extension, Potomac, and Cremorne—belong to the diagonal group and have produced some ore.

The diagonal-fracture group was the last of the major sets of fractures to form and followed as a natural consequence to continued strong shearing along the northwestward-trending group. The origin of these fissures has been clearly presented by Burbank (1933a, p. 196–200, fig. 2), and additional mapping has added no data to modify his conclusions. Briefly stated, the diagonal-fracture group consists of discontinuous tension fractures produced by the friction and pressure along the main northwestward-trending group of shear faults, which intricately divided the previously unbroken rocks between the faults of the northwestward-trending group. Figure 2 from Burbank's Arrastra Basin paper is here reproduced as figure 12, and his explanation from pages 198 and 199 is given below.

Figures 2 A and B, which illustrate the relation of the dikes near Silver Lake both before and after displacement, show that shearing stresses must have been induced in the continuous body of rock southeast of the termination of the Silver Lake dike by this shifting movement, and that rotational stress or a shifting direction of shearing force would also be induced in the footwall of the dike by the friction and pressure of the moving hanging wall block. In a body of rock subjected to nonrotational stresses, as in figure 2C, the axes of tension and compression are at right angles to each other and the planes of maximum shear occur at about 45° to these axes; but, as rotational stresses are commonly generated during actual failure, it is necessary also to consider the combination of different stresses existing in the rocks at Silver Lake and their effects upon this simple relation.

The volcanic rocks just north of Silver Lake and in the hanging wall of the Silver Lake dike are divided or sheeted by numerous essentially vertical shear planes, most of which strike about N. 25° to 40° W., but joints of another less prominent and more irregular set strike northeast. The small movements along the walls of these joint fissures and their conjugate relations to one another indicate that they were probably produced by a compressive stress applied to the rocks approximately in a north-south direction. In the area shown on plate I [Burbank's geologic map of Arrastra Basin and vicinity] only the northwesterly joints are prominent enough to be mapped in most places; for some reason the conditions favor their strong development. Figure 2E thus illustrates the relation between shear fractures and a simple compressive stress applied to bodies where the maximum relief is in an east-west direction. These relations are closely comparable with the ideal condition illustrated by figure 2C, where an axis of compression lies at right angles to an axis of tension. The shear fractures, represented by the lines S and S' in figure 2E, are represented by the sides of the parallelogram and marked by the arrows (sh). By observation it is known, however, that the planes of greatest shearing deformation in different materials vary with the properties of the material and with the rate of application of the force. Actual limits in the variation of the position of these conjugate shear joints seen north of Silver Lake are indicated by the dotted lines in figure 2 D or E. At Silver Lake the stresses that caused the conjugate joints were probably the result of radial compression induced by subsidence north of the Animas fault, and these stresses probably at the same time caused the shearing along the Silver Lake dike.

If the regional compression is combined with a shearing or rotational stress, as shown in figure 2D, the planes of maximum shearing and tensile stress deviate from their position

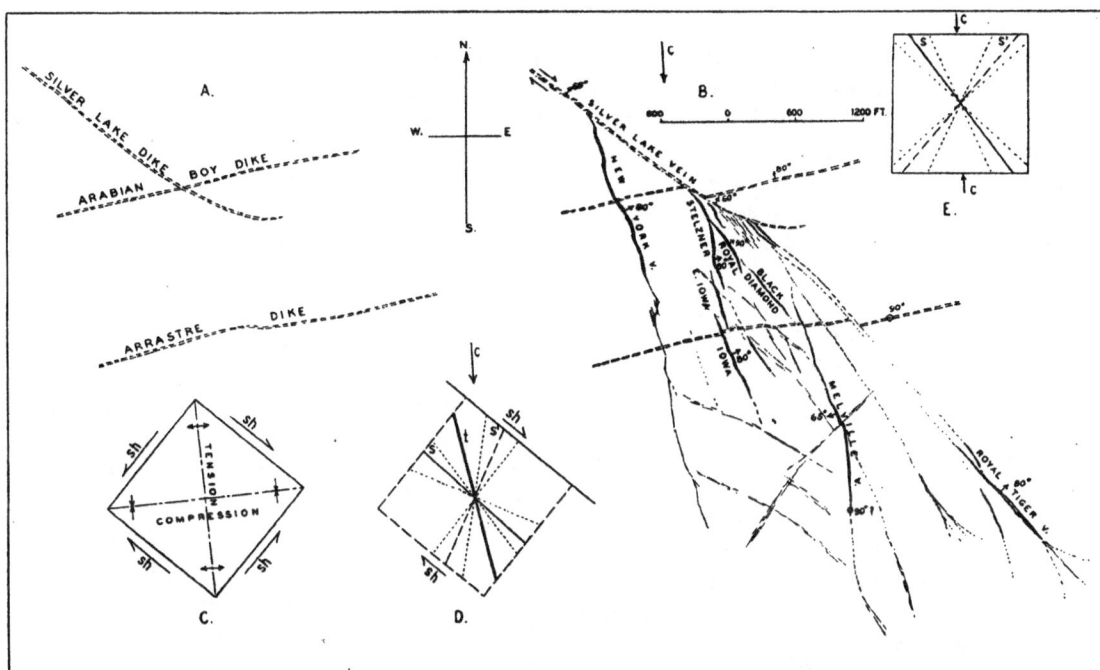

FIGURE 12.—Silver Lake vein system. A. Relation of the Silver Lake dike and other dikes before displacement. B. Relation of the Silver Lake vein system to displacement of the dikes. C. Axes of tension and compression and direction of maximum shearing stress (sh) in a body subjected to pure shear. D. Interpretation of stresses and resulting fractures near Silver Lake vein: c, regional compressive stress: sh, shearing stress: s, s', shear fractures: t, tension fracture. E. Interpretation of regional stresses in hanging wall of Silver Lake dike: s, s', shear fractures; c, regional compression. Reproduced from Burbank (1933a, fig. 2).

characteristic of the states of nonrotational shear or compression. As shown by the actual position of the footwall lodes, this is a minor change except where the planes of maximum movement are approached; here the tendency for the major fractures to become deflected toward the west is very noticeable. The change in strike and the tightening of the New York, Stelzner, and Royal veins as the Silver Lake fault is approached are described in the quotation from Ransome [1901, p. 147–148]. The weakness of the ground between the main fault and the footwall fractures at this junction and its proximity to the moving hanging wall probably account for local brecciation and the tightness of the footwall fissures.

As these footwall tension fractures, except close to the Silver Lake fissure, make a considerable angle with the plane of this major fissure and with the direction of shifting of its walls, the footwall fractures would tend to open by tension during movement along the Silver Lake fault. Furthermore, if, as appears probable, some relief from compressive stresses was afforded in an east-west direction, slight rotation of these blocks would permit the openings to be still further enlarged. Relaxation of the compression sufficient to permit settling of the hanging wall of the Silver Lake fault, which probably occurred, as in the Shenandoah-Dives, chiefly after the early epoch of horizontal shifting, would also tend to open further both shear and tension fractures in the footwall * * *.

As Burbank indicates, some rotation occurred along the slices of rock between the diagonal fractures. It was inevitable that as more fractures were produced the state of simple tension would be followed by a more complex state of stress, and some shearing would occur

along fractures that were caused by tension. For example, the Arrastra dike is displaced laterally a few feet where crossed by some of the diagonal fissures west of Silver Lake. In particular, the offset produced by the Iowa fracture (fig. 12) and the en echelon arrangement of the Stelzner, East Iowa, and Iowa veins suggest that a zone of left-lateral shearing developed about midway between the New York and Melville veins.

EASTERN SHEAR SYSTEM

The eastern shear system lies within the eastern one-third of the large triangular area described previously. (See fig. 11.) This part includes roughly the area east of Cunningham Gulch and the southern part of the area covered by this report. Most of the eastern shear system is outside the area of recent field studies. The description of its general features and the hypothesis to be presented for its origin are necessarily based primarily upon the work of others, such as the Silverton folio (Cross, Howe, and Ransome, 1905) and the later mapping by D. R. Cook and J. C. Hagen (unpublished theses; see p. A4) near Cunningham Gulch, supplemented by some personal observation. The factual data are few, and the concept of an eastern shear system should be regarded as a hypothesis to be tested in the future, and perhaps revised considerably as new relations may be discovered.

The principal faults of the eastern shear system belong to two curved sets of conjugate shear fractures, an epicycloidal set, and a hypocycloidal set, which occur only in a ring-shaped area around the southeast border of the caldera. As these geometric designations are rather meaningless, the two sets of fractures will be referred to in this discussion as the eastern arcuate group and the northwestward-trending group.

EASTERN ARCUATE (EPICYCLOIDAL) GROUP

The curving granite porphyry dikes that crop out in the southern and southeastern parts of the mapped area were previously included within the concentric system by Burbank (1933a, p. 182) and by Varnes (1948, p. 8), principally because of their dominant westward trend within the area shown on plate 1. More detailed study of their relation to regional structure, aided by unpublished data in the field notes of Cross and by more recent maps of part of the east side of Cunningham Gulch prepared by Hagen and Cook (unpublished theses; see p. A4) shows that these dikes are unrelated to the concentric system. Instead, they have been injected discontinuously into a strong shear zone that follows a curving course away from the southeastern border of the area of subsidence. The relation of this shear zone to the regional structural pattern, and its possible role as an outer limit of strong mineralization seems of great importance; but a thorough understanding of this shear zone and its dikes must await further mapping.

The dikes begin near, or within, the ring-fault zone of the caldera, 1½ miles northeast of the northeastern corner of the area shown on plate 1. Here, at an altitude of 11,400 feet on the northeast slope of Galena Mountain, Cross found a short segment of the granite porphyry dike 10 to 20 feet wide with a strike of N. 30° W. (Cross, Howe, and Ransome, 1905, p. 11; also field notes of 1901). This small outcrop of granite porphyry is not shown on the maps in the Silverton folio. The locality was visited in 1952, and several more short segments of the dike were found. A much longer exposure of granite porphyry was traced for about 1 mile obliquely up the northeast flank of Galena Mountain. The precipitous crest and the south slope of Galena Mountain and Stony Gulch were not examined. The next known outcrop of the dike to the south is near the crest of Green Mountain, southeast of the Pride of the West mine. From there the dike trends southwestward down the west side of Green Mountain through flows and tuff of the Burns quartz latite and Eureka rhyolite. Near the base of the volcanic series the dike trends due west and dips 63° N. At the contact with underlying Precambrian schist and gneiss, the dike turns abruptly southwestward, splits into two

parts, and follows rather closely the foliation of the Precambrian rocks, which here strikes N. 30° to 50° E. and dips 85° NW. The dike or dikes are hidden beneath alluvium and talus in Cunningham Gulch, but two overlapping dikes crop out on the west side of the gulch. (See fig. 5.) At the point where the dikes once more leave the schist and gneiss and enter the overlying San Juan tuff, their general course again changes from S. 40° W. to S. 70° W., although not so abruptly as on the east side of Cunningham Gulch. The dikes continue southwestward and westward and end about 2 miles east of the Animas River on the ridge north of Deer Park Creek. The total length of this arcuate group is nearly 7 miles.

Except for the last mile or two near the southwestern end, the granite porphyry dikes do not trend in a direction concentric with the caldera. The dike group does resemble, however, other curved features to the northwest of the caldera and the spiral dikes around the Stony Mountain stock near Telluride, which have been described by Burbank (1941, p. 157-161). These fractures are similar to the spiral shear lines developed when a cylindrical punch is forced into plastic material. Burbank proposed that the spiral dikes around Stony Mountain occupy shear fractures resulting from pressure directed outward from the Stony Mountain stock, and he implied that other strongly curved fissures with appreciable horizontal movement, such as the Camp Bird vein, may be shear fractures due to pressure directed outward from the central area of subsidence (Burbank, 1941, p. 214-216).

Although the curving trend of the granite porphyry dikes suggests an arcuate shear slip, the actual dike-filled fractures in the volcanic rocks west of Cunningham Gulch show no evidence of shearing movement. The dikes pinch and swell irregularly and are not connected by any strong through-going fractures. They appear to have been injected into tension fractures, of the type illustrated in figure 13, formed in the cover of volcanic rocks by left-lateral shearing in the underlying Precambrian schist and gneiss. The trend of the dikes indicate that the lateral shear movement in the schist and gneiss took place approximately along the plane of foliation, at least in the southern part of the area, where the foliation may be observed.

The belief that lateral shear movements have occurred within the Precambrian rocks along the zone of granite porphyry dikes receives some additional support from mapping by Hagen at the Green Mountain mine on the east side of Cunningham Gulch, one-half mile north of the Highland Mary mill. The southeastward-trending lower tunnels begin in Eureka rhyolite, but within a few hundred feet enter the

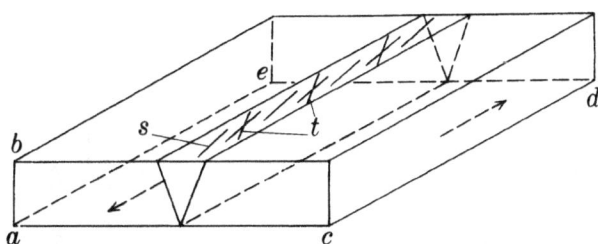

FIGURE 13.—Illustration from an experiment performed by Riedel (1929), in which a layer of clay of thickness *ab* was placed on a divided base *acde*. The two halves of the base were then displaced laterally as shown by the arrows. The resulting disturbed zone in the clay was wedge shaped. Tension cracks, *t*, and shear fractures, *s*, were produced in the clay. The tension cracks form first, and only if the clay is covered with a film of water. The en echelon shear fractures form at a lesser angle to the zone of shearing than the tension cracks and always point opposite to the direction of relative movement.

schist; a granite porphyry dike was found in the schist near the southeast end of the mine. Portalward from the dike, several faults in a zone 500 to 600 feet wide cut across the Green Mountain vein, which strikes N. 45° W. and dips 75° NE., and offset the vein a few feet to 35 feet. These faults strike north to northeast and dip steeply northwest; they are almost parallel to the foliation of the enclosing schist and gneiss and to the granite porphyry dike. The direction of relative movement, except for an apparent rotary movement of one block, is in the same relative direction along each fault; that is, the vein is displaced to the southwest in the hanging walls of the faults. Hagen interpreted the observed offsets of the vein to be due to the combination of northeast dip of the vein and normal downdip movement of the cross faults. The observed offsets of the Green Mountain vein could be due equally well, however, to left-lateral movement on the cross faults after the main period of mineralization. The direction of faulting along the eastern arcuate group would be then the same as was inferred from the en echelon arrangement of the granite porphyry dikes west of Cunningham Gulch.

Several northward-trending fractures have been included with the granite porphyry dikes in the same structural group. Among these are 2 of the 4 principal sets of fractures on Galena Mountain described by Ransome (1901, p. 171). The two sets that belong in this group strike north and N. 25° W.

The veins of Galena Mountain that trend N. 25° W. are described by Ransome as being very prominent and closely spaced nearly vertical fissures, which are filled with small stringers and in some places mineralized. The northward-striking veins are rather short and dip to the east about 80°. No important ore bodies had been found along any of them at the time of Ransome's visit in 1899. The larger lodes are accompanied by parallel fissures and veins in the country

rock. Ransome noted that one of the northward-striking veins, the Veta Madre, is cut by a metal-bearing member of the set trending N. 25° W., but he detected no offset.

Almost all the northward- and north-northwestward-trending fractures on the east side of Cunningham Gulch and on Galena Mountain are accompanied by numerous and prominent parallel fractures and joints in the wallrock. These may suggest that shear movement with a horizontal component occurred along the principal members of the sets, but no direct evidence of differential movement has been found.

NORTHWESTWARD-TRENDING (HYPOCYCLOIDAL) GROUP

This group apparently is not as prominent as the complementary eastern arcuate, or epicycloidal, group but may be represented by the Green Mountain vein, by a few fractures of somewhat irregular trend on the north end of King Solomon Mountain, and possibly by the Little Nation vein, which lies within the ring-fault zone. Right-lateral horizontal shear displacements, before and after formation of the ore, along the Green Mountain vein were noted by J. C. Hagen (unpublished thesis; see p. A4).

APPROXIMATELY RADIAL GROUP

Included in this group are several fractures that trend approximately normal to the ring-fault zone. Most of them lie outside the area of detailed study and are known only from Ransome's short descriptions and their plotted position on the folio map. There is no field evidence for lateral movement on any of them and they were excluded from the analysis of shearing in the eastern system. Among these are the prominent lodes trending N. 65° W. on the west side of Galena Mountain, which cut the veins of this area that trend northward and N. 25° W. Ransome (1901, p. 57) states, however, that no displacement was visible. There is also a prominent quartz vein, which trends northwestward along the crest of Green Mountain and which may belong to this group. It cuts an older vein that strikes N. 60° E. (Ransome, 1901, p. 170), but Ransome did not mention faulting.

The Pride of the West is the most productive vein tentatively assigned to this group. The strike of the Pride of the West vein zone (including veins worked in both the Pride of the West and Osceola mines) changes in an almost continuous arc from N. 30° E. at the south to N. 30° W. at the north. It dips 50°–70° W. At its southern extremity, the vein zone appears to have about the same strike and dip as the shear fractures associated with the granite porphyry dikes in the Green Mountain mine a short distance to the south. I agree with Cook that there has been little

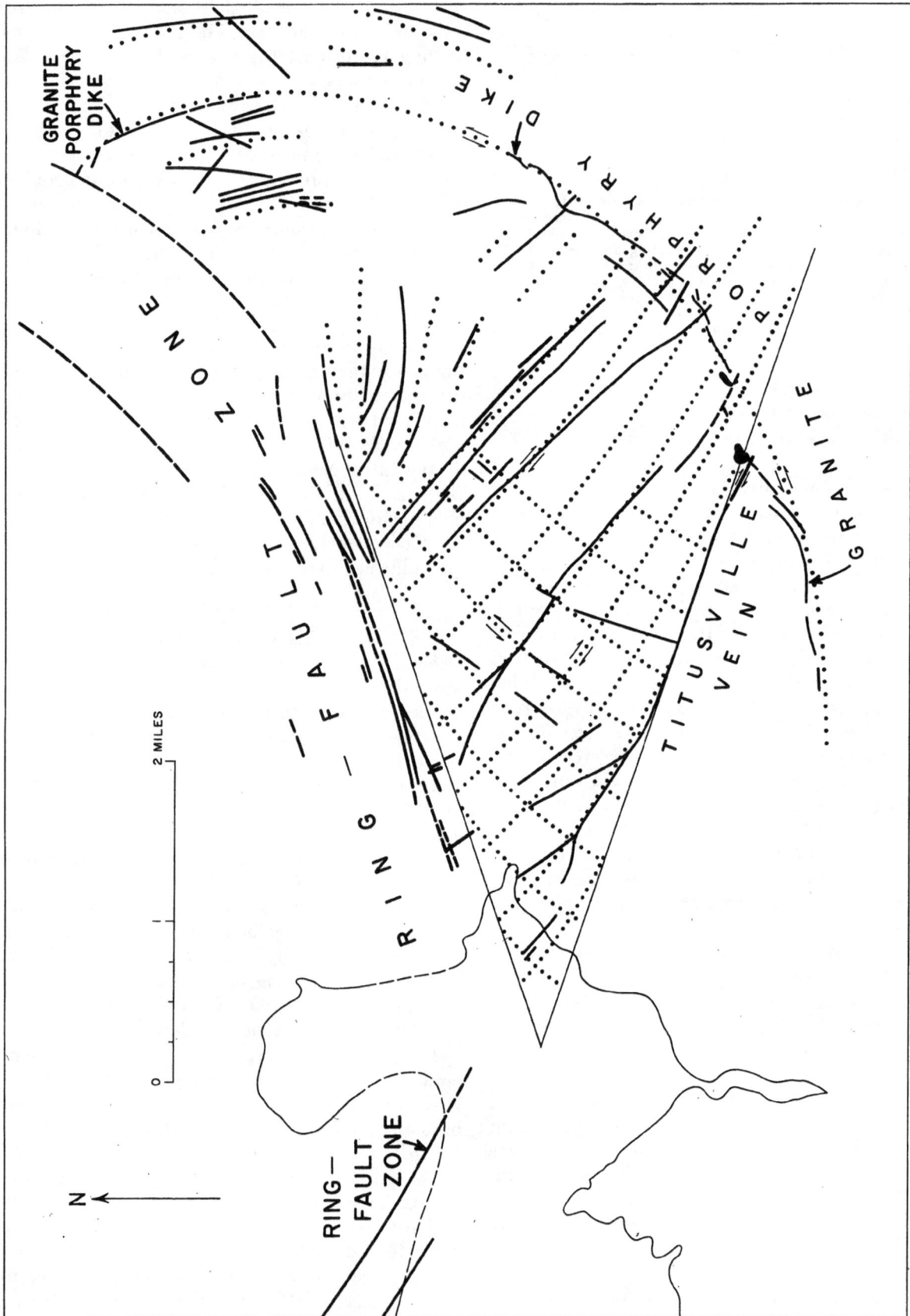

FIGURE 14.—Comparison of pattern of actual faults (solid lines) in the South Silverton district with theoretical pattern of principal shear stress trajectories (dotted lines).

or no lateral movement along the Pride of the West zone, although direct evidence for such movement would be difficult to find, owing to the complexity of the volcanic flows and prevalence of timber and talus cover.

The idealized pattern of the eastern and western shear systems together with those of the observed faults and veins is shown in figure 14. The right center part of the figure is nearly blank and corresponds apparently to an area of relatively poor mineralization between Little Giant Basin and Cunningham Gulch. Very steep slopes and cliffs made it impossible to map some of this area in detail. But exposures are fairly good and most of the few veins that were seen trend in directions having little relation to the dominant regional structure.

DIPS AND EVIDENCE FOR REVERSE FAULTING

Shear fractures of the idealized patterns shown in figure 14 are vertical; in reality the major fractures, although generally steep, dip in some places at angles as low as 55°, and it may be assumed that the departure from the idealized condition of plane strain is responsible for this difference. But it is impossible, at least at this time, to reach any general theoretical conclusions concerning the reason why the fractures dip as they do or what effect depth may have on their attitude and continuity.

The observed dips are as follows: The main northwestward-trending shear fractures of the western shear system dip northeastward, except for the Big Giant and Black Prince-Gold Lake veins at the northeast border of the group. The few data available indicate that northeastward-trending shear fractures of this system dip steeply northwestward. Northwestward-trending fractures of the eastern shear system dip northeastward, so far as is known. The dips of the Titusville fracture and of fractures of the eastern arcuate group appear to be consistently inward toward the mineralized triangular area until they approach close to the ring-fault zone, where each fracture splits into en echelon groups of fractures that dip outward. Some fractures at the northwest end of the Titusville fracture, for example the Idaho vein, dip steeply southwestward; and parts of the granite porphyry dike on Galena Mountain dip steeply northeastward. For these reversals of dip, shown diagrammatically in figure 15, there is no ready explanation.

Most of the northwestward-trending fractures within the triangular area dip for the greater part of their length obliquely toward the ring-fault zone. It might be expected, therefore, that strong compression from the north would produce reverse or thrust movements, as well as horizontal slip, along those fractures

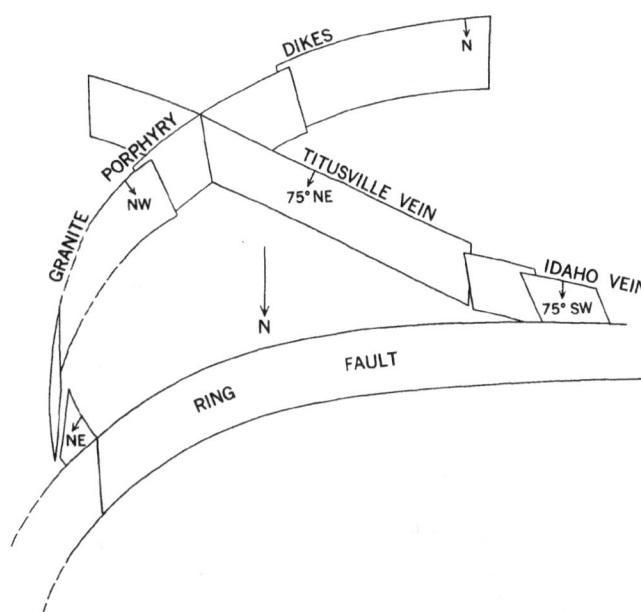

FIGURE 15.—Diagrammatic sketch of the granite porphyry dikes and the Titusville vein, as if viewed from the north, illustrating reversals of dip near the ring-fault zone.

that have a northward component of dip. Displaced contacts, however, show evidence only of normal downdip movement. Yet any thrusting would have had great influence, especially in the initial stages of deformation, upon the arrangement of individual fractures within each of the large shear zones and upon the localization of ore shoots.

Indirect evidence suggests that thrusting actually did occur along parts of the northwestward-trending group of shear fractures. In the Shenandoah-Dives vein system the principal fractures overlap as is shown, greatly simplified, in figure 16. According to the relations of overlapping shear fractures shown in figure 13, the hanging wall of the Shenandoah-Dives shear system should have moved not only southeastward but also upward relative to the footwall. Subsequent normal faulting opened what appear to be steeply dipping tension fractures, such as the Morgan fracture shown in figure 16C. The formation of the Morgan fracture between two major shears is analogous to the formation of the diagonal fractures between the Silver Lake and Titusville veins; that is, vertical section 16C is mechanically analogous to the plan view of figure 12D.

The arrangement of the major fractures in the Shenandoah-Dives vein system, which indicates initial thrusting, is reproduced on a much smaller scale by gouge planes along the Mayflower vein itself. Figure 17 shows the relation of ore shoots to changes in strike and dip and to gouge planes. It is copied from Burbank (1933a, fig. 1), except that, in the section, the initial thrust movement believed to have caused the

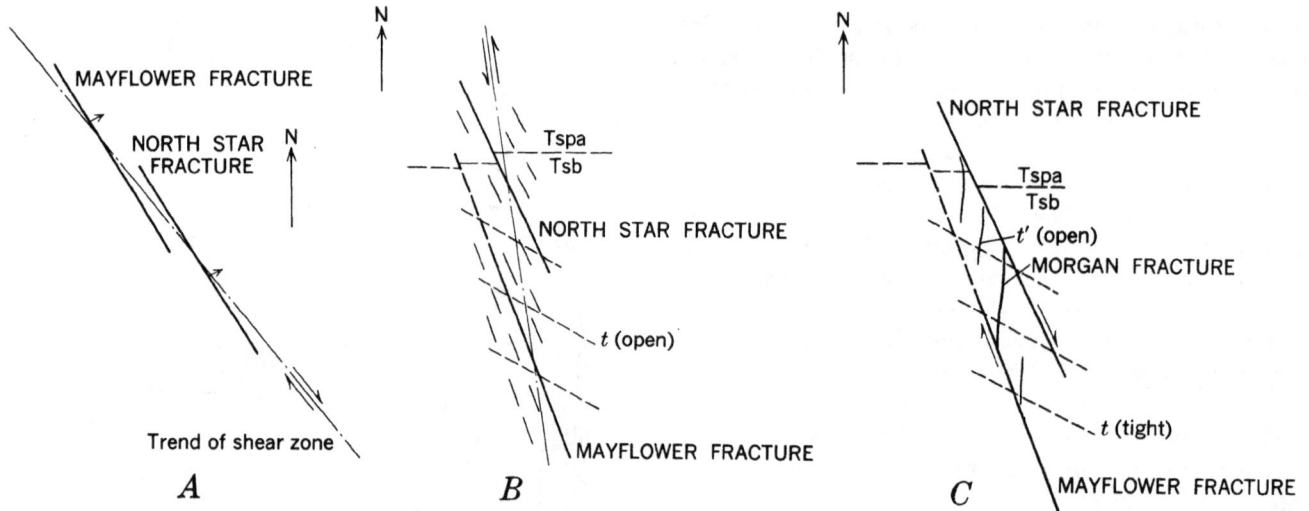

FIGURE 16.—Greatly simplified and diagrammatic relations of overlapping veins in the Shenandoah-Dives vein system. *A*, Plan showing how the Mayflower and North Star fractures trend slightly oblique to the shear zone; the obliquity is exaggerated. *B*, Section, showing en echelon arrangement of veins relative to shear zone, indicates initial thrust movement. *C*, Section showing result of normal faulting after relief of stress. Some tension fractures (dashed line), *t*, may have been opened during thrusting but became tight during normal faulting; other tension fractures (solid line), *t'*, of which the Morgan fracture is one, would open during normal faulting.

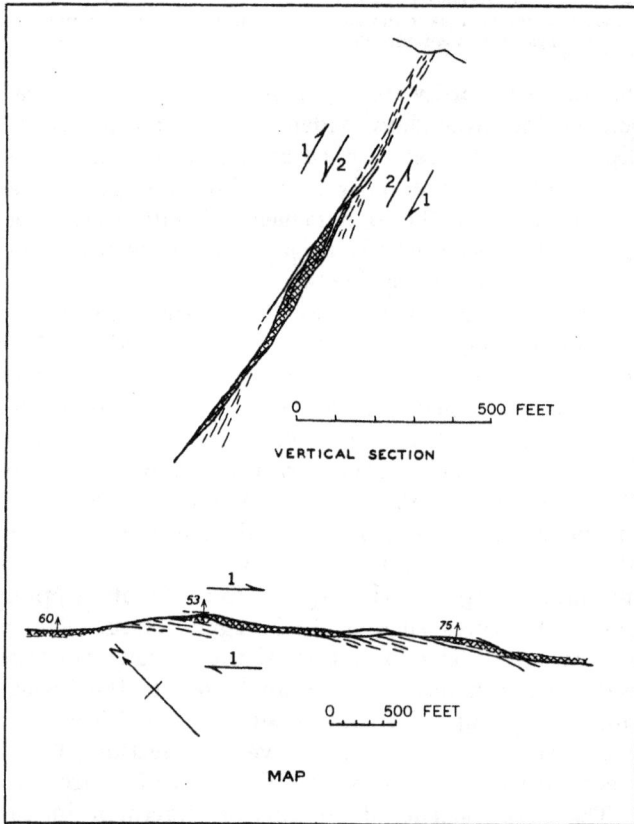

FIGURE 17.—Map and vertical section of the Shenandoah-Dives vein and ore shoots, representing diagrammatically relation of ore shoots to changes in strike and dip of veins, and to gouge planes (heavy lines). The Shenandoah-Dives dike (Mayflower dike) is omitted to simplify the diagram, and the width of ore shoots and vein are exaggerated. Ore shoots are indicated by cross hatching. Taken from Burbank (1933a, fig. 1). Arrows 1 have been added to the section to indicate the initial thrust and right-lateral movement that probably produced the gouge planes; arrows 2 indicate later normal movement.

gouge planes is indicated as well as the final normal movement. The general tendency for the gouge planes to pass, if followed upward, from the hanging wall of the fissure zone to the footwall suggests reverse faulting. According to Burbank (1933a, p. 188):

> When an ore shoot whose hanging wall is limited by such a gouge plane is stoped upward it is found that this gouge may change its dip and flatten in such a way as to pinch off the high-grade ore at the top of the shoot and then pass into the footwall.

These gouge planes are generally older than the ore, and they acted as impermeable baffles, directing the localization of ore shoots.

Strong lateral movement, accompanied by some thrusting, seems to have alternated with downward movement of the hanging walls in response to gravity. Gravitational adjustments would require that pressure from the north be relaxed somewhat. The cycle may have been repeated several times. Almost all fractures of the western shear system that were produced either by horizontal shear or by normal faulting, except the northeastward-trending group, contain vein matter in greater or lesser amounts. Therefore, at least one cycle was completed before mineralization.

Some movement has occurred after mineralization along both the northeastward- and northwestward-trending groups. Gouge-filled fissures, most of them formed after mineralization, were examined by Burbank in the Dives mine. They have striations that pitch 5°–25° NW. (Burbank, 1933a, p. 204). These suggest a thrust component in late slip. The northeastward-trending fault in Woodchuck Basin appears to offset a little the Silver Lake vein and dike and perhaps also the Titusville fracture. This fault is not conspicuously mineralized. These offsets, and their small total movement, indicate that parts of the northeastward-trending group are of very late age and may have been formed by renewed compression after the northwestward-trending fissures were so tightly sealed by dike and vein matter that failure was easier through fresh rock along the complementary direction of principal shear.

ORE DEPOSITS

HISTORY OF MINING AND PRODUCTION

For comprehensive and well-written accounts of the development of Silverton as a mining camp the reader is referred to Ransome (1901) and to Henderson (1926).

Although the discovery of gold in Arrastra Gulch first brought miners to this area in the early seventies, the discovery of silver in the base-metal ores was the major factor in establishing Silverton as a permanent camp. Between 1870 and 1890 the richer deposits were discovered and mined to the extent permitted by the prevailing methods of mining, transporting, and smelting such ores. About 1890 serious attempts to mine and concentrate the larger bodies of low-grade ore led to the successful operation of the Silver Lake mines on a large scale by E. G. Stoiber. Since 1900 the greater part of the production of the area has come from low-grade milling ores containing gold, silver, copper, lead, and zinc.

Production by principal mines through 1957 is summarized in table 6 and shown graphically in figure 18. For convenience in showing total production from the Shenandoah-Dives vein system, the production

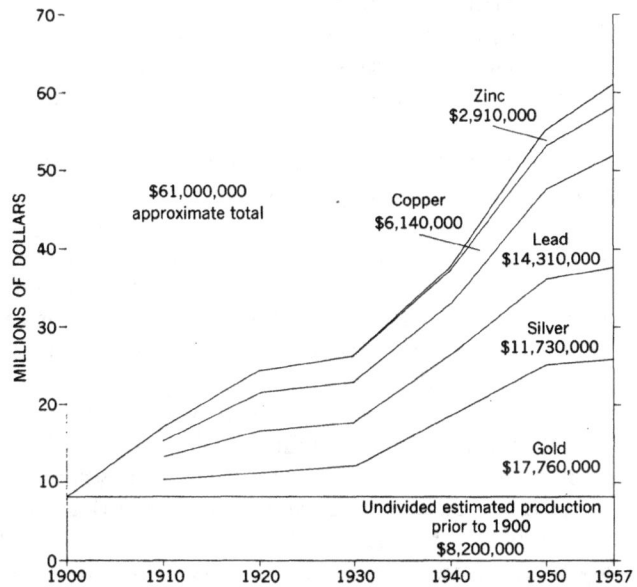

FIGURE 18.—Cumulative production record, 1900–'57, South Silverton mining area, San Juan County, Colorado. Gross value of metals in direct smelting ore or in concentrates; zinc in zinc concentrates only. Computed from average metal prices for each year, without premiums. Compiled from production records furnished by Statistics Branch, Economics Division, U.S. Bureau of Mines, Denver, Colo.

from each of several formerly independent mines, later all consolidated with the Shenandoah-Dives, has been grouped together. The production from the Shenandoah-Dives, together with that of the Highland Mary mines and Trilby Tunnel (pl. 2), has come from several individual veins within a larger fracture zone. Although the Pride of the West mine is a short distance beyond the east edge of the area shown on the geologic map, the mine was mapped and its production is also included in table 6.

TABLE 6.—Production from 1901 through 1957 of the principal mines of the South Silverton mining area in terms of total metal in direct smelting ore and concentrates

[Compiled from production records furnished by the Division of Mineral Industries, U.S. Bureau of Mines, Denver, Colo.]

Mine	Ore (thousands of tons)	Gold (thousands of oz.)	Silver (thousands of oz.)	Copper (thousands of lb.)	Lead (thousands of lb.)	Zinc (thousands of lb.) [1]	Approximate gross value [2]
Shenandoah-Dives [3]	4,099	413	8,984	22,713	65,642	19,145	$29,330,000
Silver Lake mines [4]	849	110	3,277	12,706	27,974	1,299	7,790,000
Pride of the West-Green Mountain	346	26	1,345	1,394	39,936	4,698	6,310,000
Iowa-Royal Tiger	376	46	1,210	3,982	43,043	---------	5,010,000
Highland Mary and Trilby	168	16	2,513	1,263	9,120	1	3,060,000
All others [5]	58	5	246	966	6,318	2,381	1,350,000
Total	5,896	616	17,575	43,024	192,033	27,524	$52,850,000

Estimate of approximate value of ore produced prior to 1901 [6]	8,200,000
Approximate total	$61,000,000

[1] Zinc in zinc concentrates only.
[2] Computed from average metal prices for each year.
[3] Includes production from North Star (King Solomon), Dives, Shenandoah No. 3, Mayflower, Shenandoah-Dives for 1928–53 and production from Silver Lake unit for 1944 and 1946–53.
[4] Principally production from the Silver Lake, New York, Titusville or Letter G,

Royal, and Stelzner veins.
[5] Includes Aspen, Big Giant, Black Prince, Ezra R., Gold Lake, Gray Eagle, King Solomon, Lackawanna, Little Nation, Mighty Monarch, New York, Osceola, Potomac, and White Quail.
[6] Based upon reports of the Director of the Mint and upon very general information obtained from many sources. May be too low by several million dollars.

The metalliferous deposits are of three general types: veins, replacement bodies, and disseminated deposits. Veins are by far the most important and have accounted for almost all of the production from the district.

The most productive veins have long been studied, not only for solving day-to-day problems in mining but also for scientific reasons, as instructive examples of epithermal mineral deposits controlled by geologic structures. The reader is referred again to Ransome (1901), Burbank (1933a), and the several references in the section on previous studies. No attempt is made here to repeat in detail the findings of previous workers, or even to summarize, except very briefly, the principal features of individual veins. Most of the present work was concentrated upon the structural pattern of the district as a whole. The mineralogy and paragenesis of the ores, and the finer structures of many of the veins remain largely untouched and may offer fertile fields of investigation.

VEINS

AREAL DISTRIBUTION

The section on fracturing south of the ring-fault zone already has described the pattern of veins within the district. At this point, however, it may be useful to review briefly the main systems of fractures, naming the veins that belong to each system. (See pl. 1 and fig. 11.)

First, the system of fractures that are approximately concentric with the southern border of the caldera and ring-fault zone. Most of these fractures are filled by dikes, such as the Arabian Boy and Arrastra dikes, and they are not heavily mineralized.

Second, the western shear system, which consists of two conjugate groups of shear fractures and a group of diagonal tension fractures. The northeastward-trending group of shear fractures is not mineralized, but the northwestward-trending group is heavily mineralized and includes the Titusville (Letter G), Silver Lake-Royal Tiger, Aspen, Shenandoah-Dives vein system, Big Giant, and Black Prince-Gold Lake veins. The group of diagonal tension fractures of the western shear system comprises numerous fissures trending somewhat west of north. Among these are the Scranton City vein in Swansea Gulch, the New York, Iowa, Royal, Stelzner, Black Diamond, and Melville veins of the Arrastra and Woodchuck Basins, from which has come much of the district's production, and the North Star Extension, Mountain Quail, and Potomac-Cremorne veins in Little Giant Basin.

Third, the eastern shear system, consisting of a group of arcuate granite porphyry dikes and related northward-trending fractures and veins that are not heavily mineralized, a northwestward-trending group of shear fractures that includes the Green Mountain vein, and a radial group that includes the Pride of the West vein. Lateral displacement has not been observed along members of this last group.

The ring-fault zone itself contains many veins, but most are either very narrow or barren. Strong veins where traced northward into the ring-fault zone generally weaken and are dispersed. The ring-fault zone forms an inner limit to the area of strong mineralization, at least to the area of well-defined veins. The outer limit seems to be determined at the south and east by the Titusville fracture and by the granite porphyry dikes.

MINERALOGY AND ZONING

The veins are of the complex-sulfide type and may be classed as upper mesothermal or epithermal deposits. Pyrite, galena, sphalerite, and chalcopyrite in a gangue of quartz and minor calcite are the most abundant vein minerals. Gold and argentiferous galena have accounted for most of the value. Tetrahedrite is locally abundant and often carries considerable silver. Visible free gold is very rare in this area. Ransome noted one specimen from the Royal Tiger mine, and free gold reliably is reported to have come from the Highland Mary mine. Native silver is also uncommon but was reported by Ransome from the Pride of the West and Aspen mines. Hematite, associated with chalcopyrite and pyrite in quartz and with minor amounts of galena and sphalerite, occurs in many of the veins and is especially common in the lodes of Little Giant Basin. Fluorite is abundant on some of the dumps of the Aspen mine. It was noted by Burbank (1933a, p. 204) in the Shenandoah-Dives lode in Dives Basin, where, with calcite, it forms a vein that crosses the ore-bearing fissures. Barite is a minor constituent of some veins in Arrastra Basin and in the Montana vein near Deer Park Creek; it is abundant along the southeastern parts of the Shenandoah-Dives lode. Rhodochrosite was reported by Ransome (1901, p. 73) to be common in the gangue of the Titusville vein.

Any lode generally shows some variation in mineralogy both across its exposure and along its outcrop. The variation is due in part to successive opening of individual fractures in the vein zone and to deposition of somewhat various proportions of minerals during the several stages of mineralization. However, the suite of minerals in most of the veins remains the same for several thousands of feet although the relative proportions of the minerals and the tenor of the ore may vary within wide limits.

Differences in the mineralogy of veins of one locality and those of another are neither so distinct nor so regular that zoning is well defined. Concerning zoning south of the caldera, Burbank wrote (1933a, p. 165–166):

There is a rough zonal distribution of the different kinds of ore suggesting that the igneous intrusions and the Animas fault were the principal agents controlling sources and trunk channels by which the mineralizing solutions were fed into open fissures in which ore bodies were deposited. Ore occurring in the northwesterly fissures immediately adjacent to the Animas fault zone near Arrastra Basin contain some specularite and are characterized by the base-metal sulphides, chalcopyrite, galena, and sphalerite, with a gangue of quartz and chlorite. Locally they carry sufficient free gold associated with the chalcopyrite to make it an important constituent of the ore. Ore bodies in the same fissure a mile or two farther southeast contain the base metals also, but argentiferous tetrahedrite or freibergite becomes an abundant constituent of the ores and barite, rhodochrosite, and manganiferous calcite become more abundant constituents of the gangue. Gold is here the less important and silver the more important constituent of the ore. Barren quartz and calcite succeed these silver ores still farther southeast from the fault zone. The eastern zone of the mineralized area near Maggie Gulch is characterized by extremely siliceous ores with finely disseminated sulphides and argentite. Pyrite is common in the ores. They were mined for their gold and silver content only. The ores of Sultan Mountain, at the west contain the base-metal sulphides with some tetrahedrite in a gangue of quartz and barite and in the past were mined chiefly for their silver and lead content. The most heavily mineralized and most massive base-metal veins are those of the Arrastra Basin and Silver Lake mines and the veins of Cunningham Gulch, which lie about in the center of the mineralized province south of the Animas fault. There are, however, sufficient exceptions to an idealized zonal arrangement to indicate that structure has been locally more influential than the commonly accepted temperature zones in controlling the distribution of the different ores. Furthermore, it is apparent that the largest exposed body of quartz monzonite west of Silverton does not lie above the source from which ore solutions emanated, but rather that this body and the several smaller bodies are only shallow manifestations of more deeply buried bodies of molten rock that were rising along the marginal faults and that remained partly molten long after the solidification and fracturing of intrusive rocks now exposed.

Mapping in the area surrounding Arrastra Basin has not produced many new facts to make Burbank's statements more definite. There does seem to be a general increase in calcite and barite in the gangue of the veins as they are traced southeastward, and some decrease in the proportion of chalcopyrite and hematite relative to galena. Hematite is also more abundant in the Titusville and Scranton City veins to the west of Arrastra Basin, and in the veins of Little Giant Basin to the northeast, than it is in the veins of Arrastra Basin itself.

The distribution of tetrahedrite appears to be somewhat more complex than indicated by Burbank. He noted that tetrahedrite is an important constituent in the southeastern parts of the veins, especially in the Highland Mary and Royal Tiger mines. But it was also abundant in the North Star mine on the Shenandoah-Dives lode, high on Little Giant Peak and midway along the length of the vein. More remarkable, it occurred as pockets and bunches in the ore of the Aspen mine (Ransome, 1901, p. 163) within the ring-fault zone to the northwest. Silver-rich tetrahedrite has been found in the Pride of the West, Philadelphia, and Green Mountain mines on the east side of Cunningham Gulch. On the basis of these few data, the tetrahedrite appears to be distributed in a roughly dome-shaped "shell," or zone, in the mineralized area. Much of the dome is now eroded away. The Aspen veins and Shenandoah-Dives lode trend continuously northwestward across the dome, and tetrahedrite occurs along this profile at a low altitude at the north end, at very high altitudes in the middle on Little Giant Peak, and at lower altitudes to the south in Dives Basin and in the Highland Mary mine. The rest of the "shell" would extend southward and southwestward to the higher levels of the Royal Tiger mine and eastward and downward to the lodes of Cunningham Gulch.

Most production records include the yield from several veins and from mixed ore so that the tenor of individual veins cannot always be distinguished. The available data partly bear out Burbank's statement that the ratio of silver to gold increases southeastward along the veins. The ratio is also moderately high, however, next to or within the ring-fault zone. It decreases in the vicinity of the Mayflower vein of the Shenandoah-Dives vein system and then increases sharply farther southeast. These relations are shown by some records of partial production of the several veins given in table 7.

The production records of mines along the Shenandoah-Dives vein system and its northern prolongation, from the Aspen mine southeastward to the Highland Mary mine, reflect not only the rough zoning already described but also other relations among the metals contained in the ores. Some of these relations appear to characterize the ores extracted from certain parts of the district and hence may aid in delineating zones in the district.

The production of gold, silver, copper, and lead from ores of the Aspen mine, in the years 1906 through 1929, is shown in figure 19A. In this figure the amount of each metal contained per ton of ore is plotted on an individual logarithmic scale but the scales for the several metals are shifted vertically relative to each other, so that the curves are separated. If two curves

TABLE 7.—*Silver-to-gold ratio in ore extracted from several veins*

[Computed from reports from the Director of the Mint and from records furnished by the Division of Mineral Industries, U.S. Bureau of Mines]

	Approximate ratio Ag:Au (by weight)	Period of production	Amount of ore involved in Ag:Au ratio calculation (tons)
Veins near or within the ring-fault zone:			
Idaho	100	1882	
	640	1888	
Aspen	197	1891–92	
	69	1906–30	11,600
Little Nation	300	1908–45	7,070
	275	1946–57	5,250
Lackawanna	67		1,100
Gray Eagle	70		1,200
Veins at an intermediate distance from the ring-fault zone:			
Mayflower	17	1916–19	25,800
Mayflower (Shenandoah Dives Co.)	16	1928–40	2,000,000
Silver Lake mines	31	1901–20	
	21	1941–45	
Big Giant	23		300
Little Giant	(1)	Before 1901	
Osceola	20	1946–57	30,000
Veins at a great altitude or far southeast of the ring-fault zone:			
North Star (King Solomon)	1,370	1888	
	1,343	1891	
Shenandoah and Dives claims	544	1904–08	
Highland Mary mines	2,060	1888	
	1,230	1911–40	
	41	1941–50	
Pride of the West and Green Mountain	52	to 1957	
Philadelphia	(2)	Before 1901	

¹ Low.
² Very high(?).

are parallel in such a plot the metals represented are present in a constant ratio by weight. There is a general tendency for the tenor of metals to vary together, and for the ratios between some metals to be fairly constant during several years production.

A more graphic test for covariance can be made by redrawing figure 19A so that the curve for one metal is made a horizontal line. If the curve for another metal then becomes practically horizontal, a parallelism or constant-weight ratio becomes evident. Figure 19B shows the curves rearranged so that the curve for silver is constant, or a horizontal line; each curve now represents a metal-to-silver ratio. All curves are flatter than those on figure 19A, which reaffirms the very general convariance of all metals with silver. Also, it appears that the copper-silver and lead-silver ratios vary inversely for the years 1911–27. This inverse relationship in ratios of the two metals to silver may be due to an inverse relationship between lead and copper themselves, for which there is some suggestion in figure 19A. In this case lead and copper possibly were competing for the available sulfur during deposition. The inverse ratios for silver may also indicate that copper and lead minerals were competing for silver and that these affinities differed in various parts of the ore body.

In order to explore these relationships further, similar plots were made for production from the Little

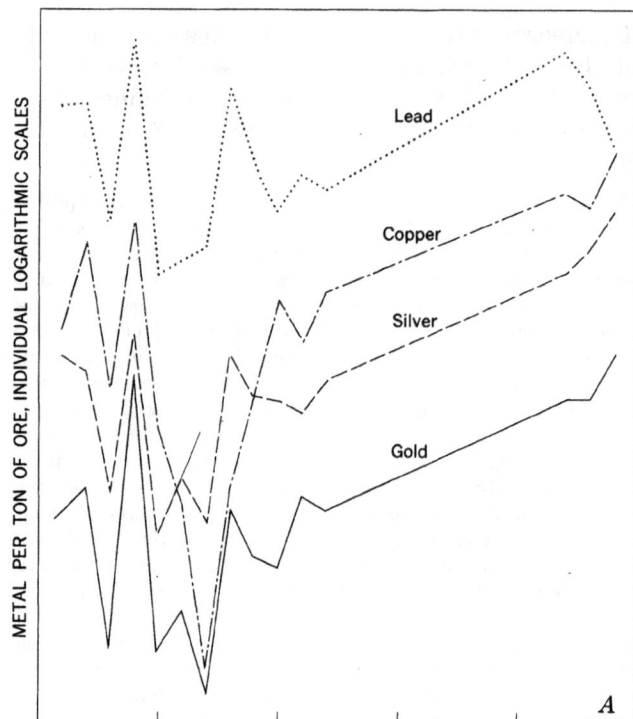

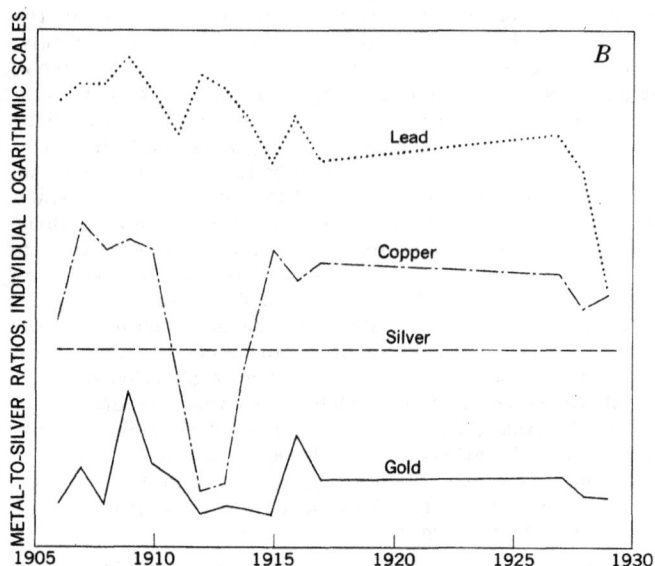

FIGURE 19.—Relations among metals contained in smelting ore or concentrations from the Aspen mine. A, Metal per ton of ore; B, Metal-to-silver ratios. Curves are to individual logarithmic scales and indicate only variations of metals relative to each other, not absolute amounts. Derived from production records furnished by the Division of Mineral Industries, U.S. Bureau of Mines, Denver, Colo.

Nation mine, which is also within the ring-fault zone in a similar geologic environment. A plot of metal contained per ton of ore, as in figure 19A, is shown in figure 20A. In this ore there is little tendency for copper to vary consistently with lead, either directly or inversely. Silver shows a fairly consistent direct variation with copper. The plot of the ratios of gold

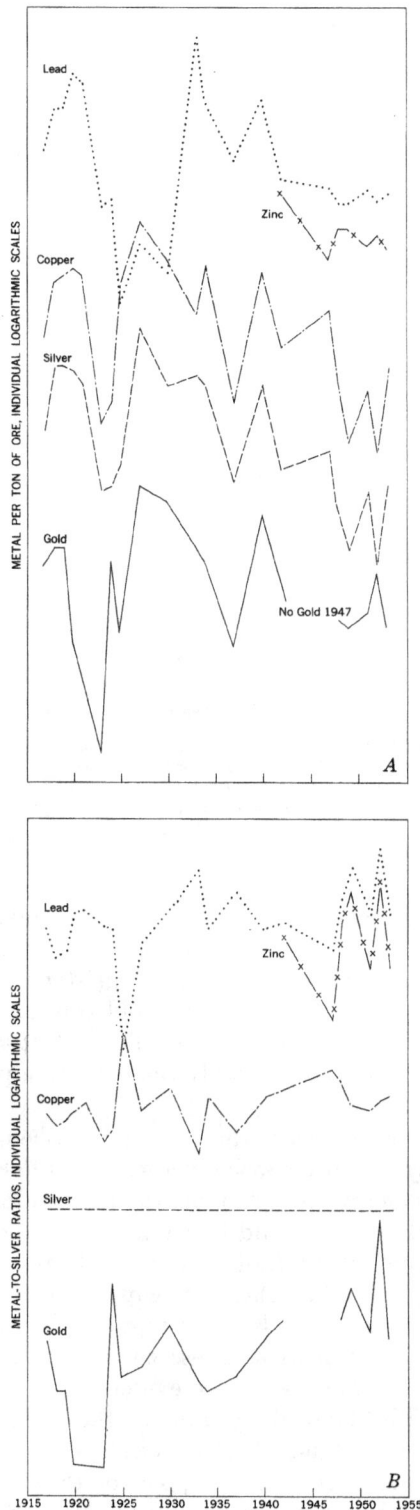

FIGURE 20.—Relations among metals contained in smelting ore or concentrates from the Little Nation mine. *A*, Metal per ton of ore; *B*, Metal-to-silver ratios. Curves are to individual logarithmic scales and indicate only variations of metals relative to each other, not absolute amounts. Derived from production records furnished by the Division of Mineral Industries, U.S. Bureau of Mines, Denver, Colo.

to silver, copper to silver, and lead to silver in figure 20*B*, to a logarithmic scale as before, now show a very striking and consistent inverse relationship between the ratios of copper to silver and of lead to silver. The average between log (Cu/Ag) and log (Pb/Ag) is very nearly constant except in very early and very late years. A truly constant relationship could be written:

$$\tfrac{1}{2}\left[\log\,(\mathrm{Cu/Ag})+\log\,(\mathrm{Pb/Ag})\right]=K$$
$$\log\left[\left(\frac{\mathrm{Cu}}{\mathrm{Ag}}\right)\left(\frac{\mathrm{Pb}}{\mathrm{Ag}}\right)\right]=2K$$
$$\frac{\mathrm{Cu}\,\mathrm{Pb}}{\mathrm{Ag}^2}=\text{a constant}$$

This equation suggests that if all metals were introduced at the same time, some chemical process tending toward equilibrium operated at the time of deposition. But without quantitative information on the mineralogy of the vein, and additional data on iron, zinc, and sulfur content of the ore, the nature of this process cannot be determined.

For ores from the area somewhat farther to the south and east the relations among the metals appear to be more complex and less consistent. This is the area of large production from the Pride of the West and Green Mountain mines, the Shenandoah-Dives operations in the Mayflower, Morgan, and other veins, and the many veins worked by the Silver Lake mines. Doubtless part of the lack of consistent relations among the metals produced from each of the several mills is due to mixing of ores from several individual veins or from widely separated stopes on a single vein. Figure 21*A* shows metal contained in concentrates per ton of ore milled by the Shenandoah-Dives Mining Co. during the years 1928-1945, excluding ore from the Silver Lake unit. There is a definite concurrence in trend between copper and silver, and to a lesser degree between copper and gold through 1941, but little or no relation holds between lead and other metals until the last years shown. Ratios of lead to silver and of copper to silver are unrelated, as shown in figure 21*B*, as long as there is a significant amount of copper in the ore.

Similar plots for ores from the Pride of the West and Green Mountain mines, not reproduced here, show a tendency for relations of all metals to be somewhat consistent until about 1936, when the relations become irregular. Ratios of copper to silver and of lead to silver vary together from 1907 to 1941, inversely from 1941 through 1951, and directly thereafter.

Silver Lake ores, which came from many veins, show no obvious relations among the metals. Content of silver appears to vary with content of both copper

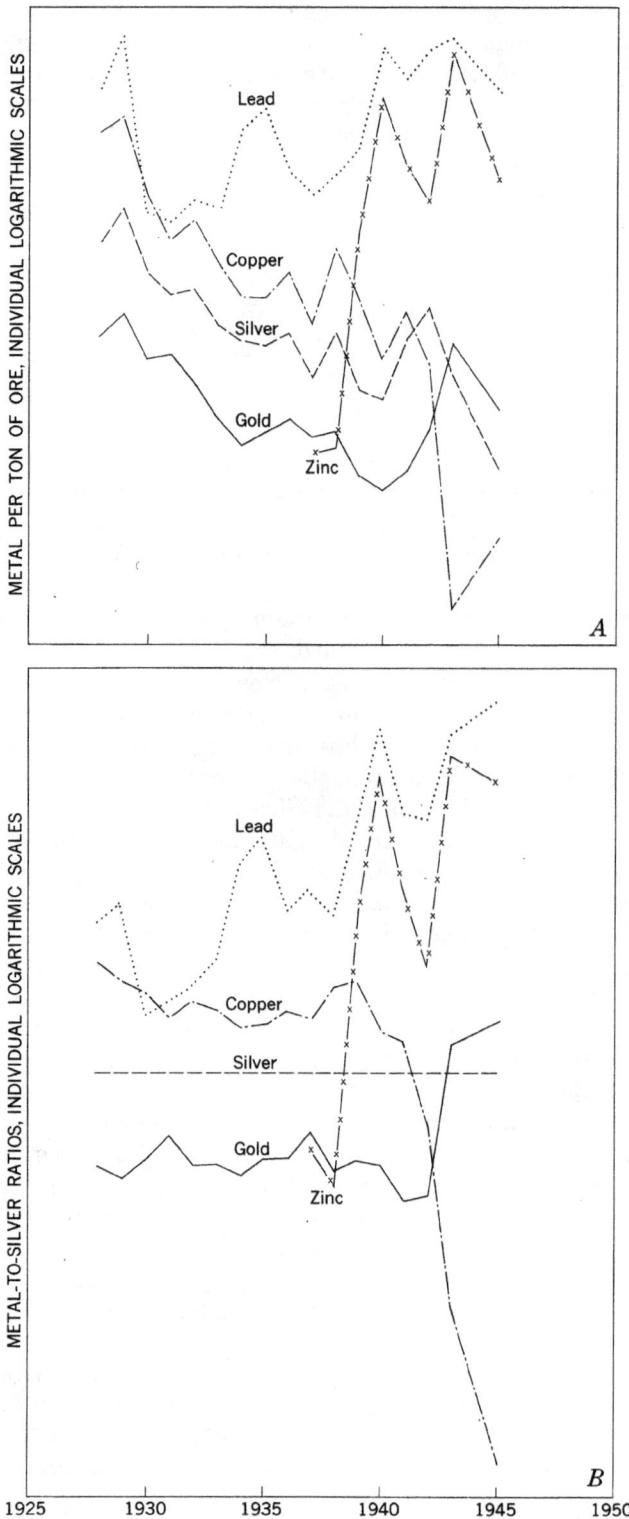

FIGURE 21.—Relations among metals contained in concentrates produced by the Shenandoah-Dives Mining Co. operations through the Arrastra Gulch portal. A, Metal per ton of ore; B, Metal-to-silver ratios. Curves are to individual logarithmic scales and indicate only variations of metals relative to each other, not absolute amounts. Derived from production-records furnished by the Division of Mineral Industries, U.S. Bureau of Mines, Denver, Colo.

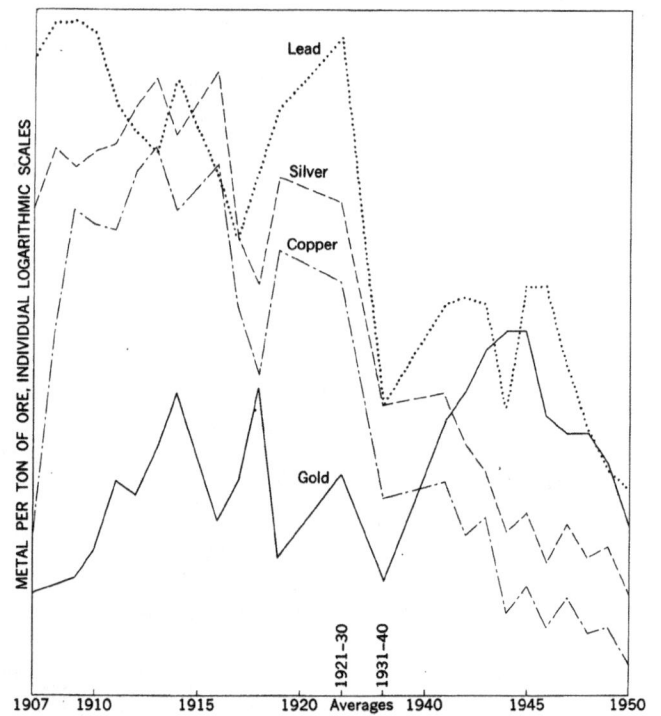

FIGURE 22.—Relations among metals contained in smelting ore or concentrates from the Highland Mary mine. Curves are to individual logarithmic scales and indicate only variations of metals relative to each other, not absolute amounts. Derived from production records furnished by the Division of Mineral Industries, U.S. Bureau of Mines, Denver, Colo.

and lead, but perhaps with lead more closely. Mixed ore from the Iowa-Tiger operations, from mines around Arrastra Basin, also shows few consistent relations among the metals. The trend of gold content follows that of silver but with smaller fluctuations. Silver appears somewhat more closely related to copper than to lead.

Still farther southeast, the ores of the Shenandoah-Dives vein system progressively assume a different and very definite character. Ore from the Dives mine shows strong correlation in trend between copper and silver, unless the content of lead is very high and that of copper very low. The content of copper tends to vary inversely with that of lead, and gold is erratic and unrelated to other metals. These changes in the character of the ore are even more evident in the production of the Highland Mary mine at the southeast extremity of the Shenandoah-Dives vein system. Contents of metals in ore or concentrates, per ton of ore shipped or milled are shown in figure 22, again to logarithmic scales that are shifted relative to each other. The remarkable parallelism between copper and silver means that the ratio of copper to silver is substantially constant. This ratio, whose average is 0.53 pound copper per ounce of silver, does not vary appreciably with the silver content, which ranges from 180 to 2.4 ounces per ton, or with annual production, which ranges from 347 to 30,400 tons of ore for the individual years shown on the graph. Some silver may be held in galena or in distinct silver minerals, but these amounts must be minor compared with the silver associated with copper minerals. Probably most of the silver is in argentian gray copper.

In summary, the broad pattern of the zoning in this area, in which there may be irregularities and exceptions, indicates that the distribution of metals in the ores appears to change similarly northward, southward, and upward from a central zone. This central zone may be represented by the deep ores in the area at the head of Arrastra Gulch, in a zone from 1 to 2 miles south of the southern border of the ring-fault zone. In this zone, gold is an important and fairly regular constituent of the ore; silver is commonly associated with both lead and copper but not dominantly with either; lead and copper either vary together or irregularly; and gray copper is not a common constituent of the ore. North of this central zone the tenor of gold generally decreases but may be locally high or very low, silver tends to associate with either lead or copper, copper and lead may vary inversely, ratios of copper to silver and lead to silver vary inversely, and gray copper is present in the ores. The same sort of changes appear to occur to the south, and in the far southeast gold again becomes low and spotty, silver varies with copper rather than lead and is probably carried mostly in gray copper, lead and copper contents are not consistently related either directly or inversely, and gray copper is an important constituent of the ore.

VEINS OF THE CONCENTRIC SYSTEM

Fractures of the concentric system do not contain veins of economic importance. Narrow zones of iron and manganese oxides, together with stringers of quartz, are common along the walls and within some of the concentric dikes. Base-metal sulfides may also be seen in some places but only in small amounts. Some of the stronger veins in the network of dikes that cross the ridge south of Kendall Gulch have been explored by pits and short tunnels. These veins contain some galena in a gangue of quartz, carbonates, and pyrite, but there is no record of ore produced. Both the Arabian Boy dike and the Arrastra dike are accompanied by narrow veins along much of their length, but there has been no work on them, except for a few prospect pits. The narrow vein or zone of altered rock that intersects Arrastra Creek at altitude 11,700 feet trends somewhat south of west, passes through the

Galicya claim, and may belong to the concentric system. Although the vein is persistent, it is apparently without economic value.

VEINS OF THE WESTERN SHEAR SYSTEM
NORTHEASTWARD-TRENDING GROUP

The northeastward-trending group of fractures is best displayed in the upper parts of Swansea and Blair Gulches and in Woodchuck Basin. Although some of the fractures here are strong and wide and greatly stained with iron oxides, only one is sufficiently mineralized to have encouraged prospecting. This one follows the northernmost of the dikes that extend across Blair Gulch and was worked by the Happy Jack mine, now abandoned. The production, if any, is not known.

NORTHWESTWARD-TRENDING GROUP

The most continuous veins of the district belong to the northwestward-trending group. They are described below in their order from southwest to northeast.

TITUSVILLE (LETTER G) VEIN

The Titusville (Letter G) vein is the longest vein in the South Silverton area. It has an average strike of about N. 71° W. and a dip of 75° NE. The structure, through most of its length, includes a compound dike, consisting of 2 or 3 adjacent sheets of andesite or latite and vein material ranging from a single filled fissure, several feet wide, to a zone of closely spaced veins as much as 100 feet wide.

The southeast end of the Titusville in Spencer Basin is a barren fault. On the western slope of Spencer Basin, the fault displaces the granite porphyry dike about 55 feet, the northeast side having moved to the southeast. The vertical movement, which may be determined nearby from offset of the top of the Eureka rhyolite, is about 35 feet, the northeast side having moved down. Northwestward over the ridge to Arrastra Basin the fault is joined by a dike, becomes mineralized, and is known in that locality as the Buckeye vein. In Arrastra Basin the vertical displacement is 80 to 100 feet. The vertical movement seems to become smaller farther to the northwest, although there are no certain means for its measurement; however, the strike-slip component increases to 350 or 400 feet, if it is assumed that segments of the Arabian Boy dike east and west of the Titusville on the south slope of Kendall Mountain were originally continuous.

Early explorations of the Titusville vein include the Buckeye mine to the southeast, in Arrastra Basin, and several pits and short tunnels to the northwest, on the south slope of Kendall Mountain. Ransome reported

(1901, p. 161) that the ore of the Buckeye consisted of "chiefly galena, with some tetrahedrite, the latter being more abundant in the upper tunnel." To the northwest of the Titusville mine, along the north side of Kendall Gulch, the vein is 3 to 10 feet wide and, judging from the poorly mineralized outcrops, appears to be relatively barren. The course of the vein changes abruptly to a more northward direction near the mine at altitude 12,504 feet, then returns to its general northwesterly trend at the crest of the ridge east of Kendall No. 2 Peak. Where it trends northward, the vein zone is unusually wide, about 100 feet, and contains at least two dikes within a swarm of quartz veins. Outcrops of some of the veins show galena and other sulfides and have been explored by trenches. The abnormal widening of the vein zone appears to be the combined result of the movement of the hanging wall southeastward relative to the footwall and the change in strike.

Mining on the Titusville vein began near the divide between Kendall Gulch and Arrastra Basin. The Titusville mine was operated from 1886 until 1893. According to Ransome (1901, p. 161) the mine—

was developed on four levels, aggregating about 1,150 feet of drifts. The lode is about 30 feet wide. Its general dip is to the northeast at about 80°, although it is locally vertical. It consists of chalcopyrite, galena, sphalerite, and pyrite, in a gangue of quartz and rhodochrosite. Chalcopyrite is the most abundant ore mineral. The ore is generally of low grade, that containing abundant chalcopyrite running about $6 per ton. There is a richer streak, however, usually about 14 inches wide, which may run as high as $30 per ton. Some of this richer ore contains very finely crystalline galena scattered through the quartz in minute particles. Where this occurs the ore may carry 1 to 3 ounces of gold and 60 to 100 ounces of silver per ton, the value being associated with the presence of galena.

The most extensive work on the Titusville vein was done from the Silver Lake mines through deep-level crosscuts driven southward from the New York vein. Much ore was mined in the period 1904–06 both from under the southwest side of Arrastra Basin and from under Kendall Mountain to the northwest. The ore was milled with ore derived from other veins, so that there is no clear record of the quantity or tenor of ore mined solely from the Titusville vein. Both the high-level workings of the Titusville mine and the deep drifts of the Silver Lake mines were inaccessible during the 1945–46 field seasons.

Between July 1953 and May 1954, the Titusville vein at the level of the Silver Lake crosscut from the Shenandoah-Dives mine was explored under an agreement between the Shenandoah-Dives Mining Co. and the Defense Minerals Exploration Administration (DMEA 2991). The 4,177-foot drift along

the Melville vein was extended southward as a crosscut to the Titusville and for about 1,130 feet northwestward and 490 feet southeastward along the vein. The work was supplemented by more than 1,200 feet of diamond drilling. The mineralized rock found by the exploration was of very low grade and work was discontinued before the exploration program originally contemplated was finished.

The Titusville dike, or dikes, ends near the crest of Kendall No. 2 Peak. A pile of gossan lies beside a prospect pit just below and to the west of the crest of the peak. Galena in the cores of many of the fragments is intimately cut by microscopic seams of anglesite or cerussite. Still farther to the northwest the vein zone is less continuous and consists of several overlapping mineralized fissures. No serious exploration was attempted on the northwest side of Kendall No .2 Peak except at the Idaho mine. The Idaho mine is now inaccessible, but according to Ransome (1901, p. 161) the ore was mostly chalcopyrite and pyrite, with pockets of galena in a quartz gangue. Such ore may be seen on the present dump, but probably much of it has been mined since Ransome's visit.

SILVER LAKE VEIN

The average strike of the Silver Lake vein is N. 57° W. and the average dip 60° NE., which is unusually low for a vein of the northwestward-trending group. The northwest end of the vein is concealed by till and timber on the south slope of the Animas Valley between Swansea and Blair Gulches. The vein is well exposed on the crest of the ridge west of Blair Gulch, but the outcrop does not become strong and continuous until the vein passes southeastward into Woodchuck Basin. Over Round Mountain and down to Silver Lake, the vein, where not covered by talus, is marked by a prominent gossan-stained zone as much as 40 feet wide. From the farthest northwest exposure nearly to Silver Lake the vein follows the footwall of an andesite or latite dike. The dike turns to the east near Silver Lake and ends, but the vein apparently continues to and beneath Silver Lake in several separate fissures (fig. 23).

The net differential movement between the walls of the Silver Lake fissure is in the same direction as those of the Titusville and other northwestward-trending fractures, the northeast side having moved down and to the southeast relative to the southwest side. Most of the movement took place before vein deposition. Burbank estimated the net amounts of displacement near the Arrastra and Arabian Boy dikes to be about 200 feet horizontally and about 100–150 feet vertically (1933a, p. 191–192). The vertical displace-

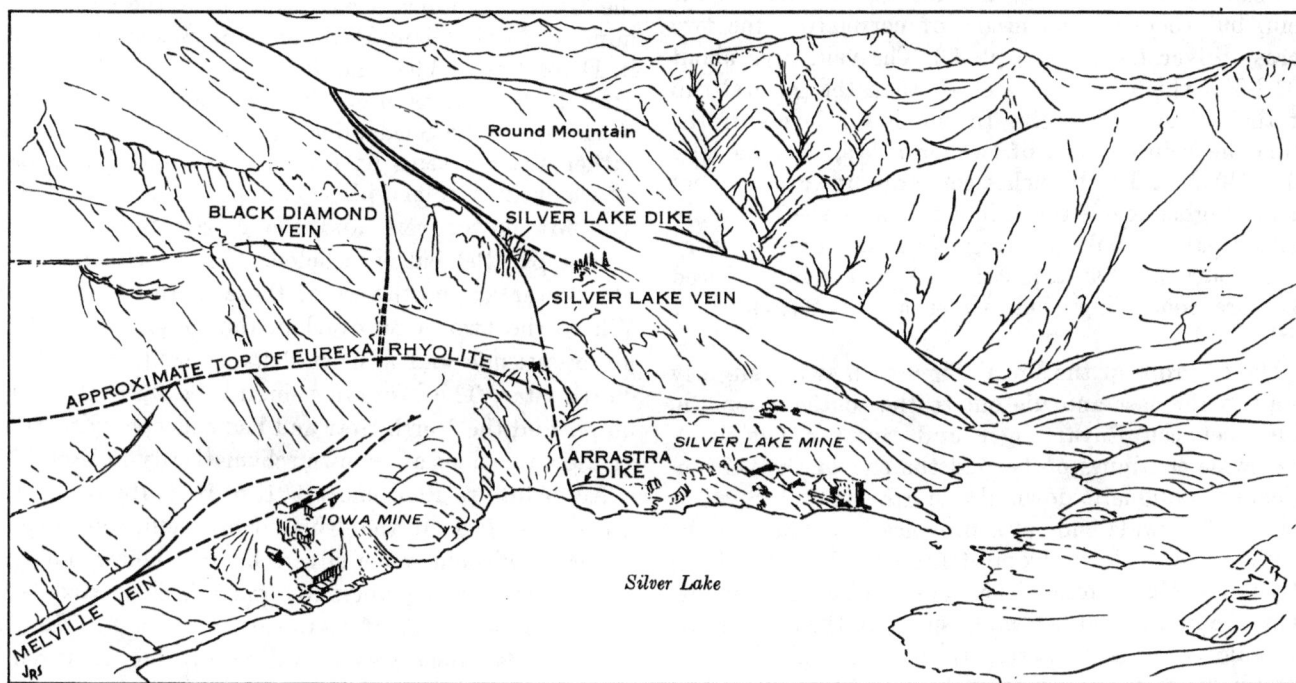

FIGURE 23.—Sketch and view northwestward across Silver Lake. Silver Lake fault, dike, and vein pass through the saddle to the left of Round Mountain.

ment where the vein crosses the ridge west of Blair Gulch is about 250 feet, but the horizontal displacement could not be measured. Some movement almost parallel with the vein took place after the main period of mineralization, as shown by gouge zones on the walls and broken vein matter. According to Ransome

(1901, p. 152), the movement after deposition of the ore was, in the main, oscillatory, with little net displacement.

The more productive parts of the Silver Lake vein lie beneath Woodchuck Basin, Round Mountain, and Arrastra Basin. The Nevada mine explored the vein beneath the ridge between Woodchuck Basin and Blair Gulch. A drift at the Unity level was run northwestward beneath the old Nevada workings, but stope maps indicate that not much ore was removed from these parts of the vein. The workings are much more extensive beneath Round Mountain. Here there are many stopes between 12,000 and 12,500 feet altitude, operated by Silver Lake mines at an early date, and, later, comparable amount of stoping at lower levels done by the Shenandoah-Dives Mining Co. via the Silver Lake crosscut.

The vein ranges in width from a few feet to 20 feet. The product has been dominantly a low-grade lead-silver ore, and the principal minerals are pyrite, galena, sphalerite, chalcopyrite, and gray copper in quartz gangue. There is very little gold in most of the ore. The upper parts of the vein contained much broken material and products of alteration.

ROYAL TIGER VEIN

The Royal Tiger vein crops out on the southeastern slopes of Arrastra Basin. It may be a continuation of the same general zone of fissures as the Silver Lake vein, but there are no means of correlating the two across Silver Lake. (See pl. 1.) The vein dips about 80° NE., which is much steeper than the general dip of the Silver Lake vein. In the lower levels of the mine the vein consists of two branches, separated by 30 or 40 feet. The branches converge upward, so that in the higher levels the lode consists of the two main veins separated only by irregular stringers of country rock (Burbank, 1933a, p. 201–202). The vein produced lead ore containing some silver and, occasionally, a little gold.

The outcrops of the Royal Tiger vein swing slightly more to the east and die out to the southeast on the ridge between Royal Tiger and Spencer Basins. A low swale continues on beyond the end of the vein on a course obliquely down the north side of Spencer Basin. The swale ends at a prevolcanism fault on the south side of a down-dropped fault block south of the Highland Mary mine. The rough alinement of the vein, the swale, and the fault suggests that where the volcanic cover is relatively thin, as in this locality, fracturing of the volcanic rocks may be controlled to some degree by old faults within the underlying Precambrian rocks.

SHENANDOAH-DIVES VEIN SYSTEM

The Shenandoah-Dives vein system is one of the major faults of the western shear system and furnished the major production of the district for many years. It may be traced from Arrastra Gulch southeastward over the high ridges and deep basins nearly to Mountaineer Creek, a distance of about 12,000 feet. Its average strike is N. 40° to 50° W. and the dip is 60° to 80° NE. Its structure is complex; several rather distinct but closely spaced veins have been mined within the zone.

The northwest end of the Shenandoah-Dives vein system is hidden by talus in Arrastra Gulch (fig. 24), but the veins of the Aspen mine, west of Arrastra Gulch, may possibly be a northwestward continuation of the system. These veins are overlapping and discontinuous, at least on the surface, and it is hardly possible that any one of them could be followed underground to the Mayflower vein of the Shenandoah-Dives mine. According to Ransome (1901, p. 163), the ores of the Aspen mine consisted of galena, sphalerite, chalcopyrite, pyrite, tetrahedrite, and a little native silver in the higher levels. The gangue is quartz with considerable amounts of green fluorite. The mine has been idle for many years, and most of the adits are caved.

That part of the vein system extending from Arrastra Gulch to Dives Basin was first explored by the Mayflower mine in Arrastra Gulch, the North Star mine on Little Giant Peak, and the Shenandoah mine in Dives Basin. These holdings were consolidated, and after 1926 the three mines were connected by extensive workings of the Shenandoah-Dives Mining Co. Three rather distinct veins have been mined along this part of the vein system (pl. 3 and pl. 1, section A–A'). The Mayflower vein and the North Star vein are almost parallel but en echelon. The ore shoots of the Mayflower are northwest of those in the North Star. Where the two veins overlap, the Mayflower is at a lower altitude and a short distance southwest of the North Star. The Morgan vein lies between the Mayflower and the North Star and has a steeper dip. Some of these relations are shown schematically in figure 16.

According to Ransome (1901, p. 164), the ore in the upper 200 feet of the North Star vein was largely anglesite without much silver. The ore below the zone of oxidation was principally galena, associated with increasing amounts of tetrahedrite at depth. In the lower levels galena was subordinate to highly argentiferous tetrahedrite. Mining had ceased before 1900 at about the fifth level of the North Star mine (approx alt 12,750 ft). These high parts of the vein were not explored again until the Shenandoah-Dives Mining Co. drove a raise from an altitude of about 11,220 feet directly beneath the crest of the ridge. The deep-level mining development showed that the Mayflower vein contained abundant chalcopyrite in the lower levels; the vein was mined chiefly for gold, but also for silver,

FIGURE 24.—Arrastra Gulch viewed from the north side of the Animas Valley. Shenandoah-Dives upper tram terminal, bunkhouse, and main portal are below cliffs of Burns quartz latite and Eureka rhyolite at left center. The Shenandoah-Dives vein is in the dark ravine to the left of the bunkhouse. (See also fig. 3.) Location of Aspen mine is indicated by dumps in timber at lower right. The southern boundary of the ring-fault zone crosses the ridge at the right about at timberline; the faults are marked by shallow east-west swales across the ridge.

copper, and lead (Chase, 1929). The Morgan vein was mined also chiefly for gold.

The Shenandoah-Dives vein system continues southeastward from Dives Basin, but apparently splits into several parts as it crosses the ridge on the north side of Royal Tiger Basin and finally dies out in the pre-Tertiary fault-block and landslide area near Mountaineer Creek. This part of the vein has been mined intermittently from the earliest days of the camp, and for many years was worked by the Highland Mary Mining Co. The ore here is principally galena, chalcopyrite, and pyrite, with some sphalerite, tetrahedrite, and a little free gold in a quartz and calcite gangue.

The Shenandoah-Dives vein system follows an andesite dike, called the Mayflower dike, along the surface from Arrastra Gulch southeastward up the slope to an altitude of about 12,600 feet, where the dike ends.

This dike has been identified on the main Shenandoah-Dives level as far southeast as the 4500 mine coordinate, a little southeast of Little Giant Peak.

According to Burbank (1933a, p. 186) the net postdike displacements along the Shenandoah-Dives vein system are from 125 to 140 feet vertically and about the same horizontally. The vertical displacements southeast of Dives Basin are very small or nil and there are no means of estimating horizontal movements. Where they can be measured, the net displacements along the Shenandoah-Dives vein system are in the same direction as those along other northwestward-trending fractures of the western shear system, the northeast side having moved to the southeast and the hanging wall down.

The control exerted by these movements and by the dike upon fissuring and localization of ore shoots has been described in detail by Burbank (1933a, p. 185–

191, 202–212) whose main points are summarized in the following statements: Where the vein system followed closely along a preexisting dike wall, northwest of Little Giant Peak, the various types of country rock and the manner in which they fractured in response to stress had little influence upon the structure of the vein system. Here the openings for vein material seem to have been provided by irregularities in the dike wall that formed openings during differential movements of the walls. Movement of the ore solutions appears to have been controlled by gougy impermeable slip planes that commonly pass gradually from one wall of the dike to the other when traced horizontally, and from the hanging wall to the footwall when traced upward. Fracturing along the southeastern part of the lode where there is no dike appears, on the other hand, to have been controlled more by the strength and other physical properties of the wallrocks. In brittle rocks, such as massive dense flows, the lode is dispersed along numerous fractures; in tough or plastic rocks, such as some of the volcanic breccias, the lode more commonly follows along a single fracture.

NORTHWESTWARD-TRENDING VEINS OF LITTLE GIANT BASIN

The principal northwestward-trending veins of Little Giant Basin are the Big Giant vein and Black Prince-Gold Lake vein, a few hundred feet to the northeast. These are prominent veins related in origin to the Silver Lake, Shenandoah-Dives, and other members of the northwestward-trending group, but they have never been very productive. They trend about N. 45° W. within Little Giant Basin but swerve more westward on passing down into Arrastra Basin. Much of this curvature in outcrop results from the southwestward dip of the veins, which is the opposite of the usual dip of northwestward-trending veins. The dip of the Big Giant vein is 70° to 80° SW, and that of the Black Prince-Gold Lake 55° to 80° SW. No displacement was observed along the Big Giant vein. Vertical displacement along the Black Prince-Gold Lake fissure was determined by Burbank at the top of the Eureka rhyolite near the lip of the basin to be about 175 feet, with the hanging wall having moved down. I estimate the vertical displacement near Gold Lake to be 110 feet. The vertical displacement decreases to the southeast or is distributed irregularly along many fractures, for the base of the pyroxene-quartz latite on King Solomon Mountain is at about the same altitude as it is on Little Giant Peak on the other side of the Gold Lake vein. No means were available for determining horizontal displacements.

Quartz, chalcopyrite, and hematite appear to be the chief constituents of the Big Giant vein. Some stopes were opened on the vein but the mine has not been worked since before 1900. The Black Prince vein was explored for about 1,000 feet by an adit near the small lake in the lower part of the basin, and a little ore was removed. The King Solomon vein on the east side of the basin shows galena, chalcopyrite, and pyrite at the surface and has been mined from three adits, but the amount of ore produced is not known.

The King Solomon vein forms the northeast edge of the northwestward-trending group of productive veins in the western shear system. Farther east and northeast there are a few quartz-calcite veins that carry little or no sulfides. The area between Little Giant Basin and Cunningham Gulch appears, as mentioned before, to be a zone of relatively barren veins between the productive veins of the eastern and western shear systems.

DIAGONAL GROUP

SCRANTON CITY VEIN

The Scranton City vein diverges northward from the hanging wall of the Titusville vein, approximately at the crest of Kendall Mountain. The inclusion of the Scranton City in the diagonal group is somewhat questionable, because the northern part of the fissure seems to follow an andesite dike. All other diagonal fissures in the area appear to have been opened after the period of intrusion of andesite dikes. The trend of the vein in relation to that of the Titusville and the lack of evidence of strong differential movements indicate, however, that the Scranton City vein is primarily a tension fracture that was opened by shear movements along the Titusville vein; hence, it is more properly included in the diagonal group.

Where seen in the higher workings, altitude 12,393, the lode consists of 2 or 3 veins within a mineralized zone about 30 feet wide that contains quartz, chalcopyrite, pyrite, and minor amounts of sphalerite. The veins strike N. 15° W. and dip about 77° E. The vein zone is covered by talus through much of its course, but presumably continues northward along the west side of Swansea Gulch and becomes dispersed in the ring-fault zone near the Lackawanna mine.

SILVER LAKE FOOTWALL VEINS

The veins that trend somewhat west of north in the footwall of the Silver Lake vein in Woodchuck and Arrastra Basins have been much more productive than the Silver Lake vein itself, and they have probably accounted for about one-fourth of the total value of ores mined in the South Silverton area. Many of them are prominent veins that were discovered and worked during an early stage in the mining history of the region, but the large-scale operations necessary for profitable mining of the low-grade ore were not started until the erection of the first Silver Lake mill by E. G. Stoiber about 1890. The many claims were gradually consolidated by Mr. Stoiber and his wife and were sold in 1901 to the American Smelting and Refining Co., under whose ownership the Silver Lake mines continued to be leading producers in the area for about 15 years. The Iowa-Tiger group of claims, adjoining the Silver Lake group, also produced large tonnages of ore until about 1920. Development of the Silver Lake footwall veins has been carried on since 1941 at deeper levels through the Silver Lake crosscut from the Shenandoah-Dives mine, under a leasing agreement with the American Smelting and Refining Co.

The arrangement of the Silver Lake footwall veins may be seen in plate 1 and more clearly in figure 12. The relations change with depth, so that some veins,

FIGURE 25.—View of south face of Little Giant Peak. Light-colored area in notch below the peak is the oxidized surface of the Shenandoah-Dives vein system. Diagonal fractures in the footwall of the Shenandoah-Dives fault traverse the cliffs in the foreground.

which are separate at the surface and in the higher levels, join at lower levels. At the level of the Silver Lake crosscut, for example, the Iowa and East Iowa veins are apparently one vein; similarly, the Royal, Black Diamond, and Melville veins here cannot be differentiated. Another vein that is included in the Silver Lake footwall veins but that does not crop out was found at the Silver Lake crosscut between the Silver Lake vein and the Stelzner vein, 3,490 feet from the turnoff from the Shenandoah-Dives main level.

DIAGONAL VEINS OF LITTLE GIANT AND DIVES BASINS

Several veins diverge southward from the Big Giant and Black Prince-Gold Lake veins in Little Giant Basin. They originated in the same way as the diagonal veins south of the Silver Lake vein, but they are much smaller and almost barren. Only the Mountain Quail, North Star Extension, and Potomac-Cremorne veins have been explored.

Fractures of the diagonal group in the footwall of the Shenandoah-Dives vein system show prominently in the cliffs at the head of Dives Basin south of Little Giant Peak (fig. 25). They do not appear to be mineralized and there are no mines on them.

VEINS OF THE EASTERN SHEAR SYSTEM

Most veins of the eastern shear system lie east of Cunningham Gulch, beyond the area mapped in this study, and were not examined except at the Pride of the West and Green Mountain mines. The relations of the many veins of this area to the regional structure are therefore uncertain. The Green Mountain and Little Nation veins are included in the northwestward-trending group of the eastern shear system, and the Pride of the West vein zone is tentatively included in the radial group. The Green Mountain mine was visited briefly but not mapped. Excellent maps and a discussion of the ore deposits and local geologic setting of this mine have been prepared by John C. Hagen.

LITTLE NATION VEIN

The Little Nation, or Royal Charter, mine is one-half mile southwest of Howardsville and 700 feet above the Animas River. This mine is one of the few that have been active in the ring-fault zone, on the southern border of the subsided block, and has a recorded production of about 12,300 tons of lead-silver ore. The workings follow several discontinuous overlapping veins that trend about N. 50° W. (pl. 4). The veins have been faulted a few inches by northeastward-trending fractures, probably during postmineralization movement within the ring-fault zone.

PRIDE OF THE WEST VEIN

The Pride of the West mine is just beyond the east edge of the area of the topographic map, on the east side of Cunningham Gulch. It was mapped with the assistance of George Sowers in 1945 and revisited briefly in 1946. Since then, a very thorough study of the Pride of the West vein and the Osceola vein, which is an adjoining property to the north, has been made by Douglas Cook.

The workings are extensive and more complex than most mines in the district, owing to the irregularity of the vein zone. The veins trend roughly north but curve somewhat convexly to the east (pl. 5). The dips are generally 50° to 70° W. Ore is localized along veins within a fracture zone as much as 100 feet wide. Normal faulting apparently took place along this fracture zone, and as a result tension fissures opened between individual faults. The tension fissures generally dip at steeper angles than the faults themselves and were favorable sites for subsequent mineral deposition. The faults were also mineralized, locally, particularly in the southern part of the mine. Some movement along the faults continued during and after mineral deposition.

Sections $A-A'$ to $I-I'$ across the mine workings (pl. 6) show, more clearly than the plan map, the relation of ore shoots to the faults. Section $C-C'$ is typical and shows relatively flat faults and gouge seams with ore-bearing veins between the faults.

The steeper veins are generally wider, more continuous, and richer, especially in the northern part of the mine. (Note the Lane vein on plate 6, sections $A-A'$ and $B-B'$.) The veins split and it is difficult to trace them in the middle part of the mine (between sections $D-D'$ and $G-G'$). This area is transitional between the southern part of the mine, where the faults are themselves mineralized, and the northern part, where only the steep fractures between the faults are strongly mineralized.

Gold is generally fairly low and erratic in occurrence. Occasionally some wire silver is found, and the silver content is generally high wherever gray copper is present. Galena, sphalerite, chalcopyrite, and pyrite are the chief sulfide minerals, and quartz and calcite are the principal gangue minerals. Banding, cross-cutting veinlets, and breccia cemented by vein matter indicate that ore deposition took place in several stages and was accompanied by movement of the vein walls. In many places the veins have no true walls, but rather a false hanging wall or footwall that is only a thin slab of barren rock which conceals another vein on the other side. Because of the discontinuity and splitting habit of the veins, the operation of the mine requires attentive supervision.

OXIDATION OF VEIN MINERALS

Sulfide minerals within the veins of the South Silverton area are very little oxidized at altitudes below the upper limit of glaciation; galena, sphalerite, and chalcopyrite may be seen at the surface with only a film of discoloration. Above the upper cutting limit of the glaciers, however, the veins are commonly stained and oxidized to depths of 100 to 200 feet. The upper cutting limit of the ice is probably at or a little above the top of the highest line of cliffs along the sides of the valleys. This line drops in altitude from the heads of the cirques toward the mouths of the valleys. Above this line, the oxidized material produced by weathering in preglacial or interglacial periods has not been removed. Anglesite was abundant in the highest levels (17th and 19th) of the Shenandoah-Dives mine, and the soft decomposed vein matter required special techniques for mining and milling. Anglesite is also abundant on dumps along the Titusville vein at an altitude of about 13,000 feet on the west side of Kendall No. 2 Peak.

INFLUENCE OF WALLROCK

Additional mapping bears out Burbank's (1933a, p. 204-211) conclusion that, except where veins follow preexisting dike-filled fractures, the vein material is distributed along more numerous and smaller fractures in brittle rocks, such as massive flows, than in tougher more plastic volcanic breccia. Contrary to the prevalent interpretation of Burbank's remarks, however, the Eureka rhyolite is by no means everywhere a less favorable wallrock than the Burns quartz latite, for both formations contain both tough and brittle rocks. In any event, if a vein follows a dike, the dike wall exerts more control over the fractures filled by the vein material than does the physical character of the wallrock.

Tuff-breccia of the San Juan tuff has proved to be a favorable wallrock for the Shenandoah-Dives vein system at the Highland Mary mine. (See pl. 3.) The vein system is also strong and continuous at this locality in the Precambrian schist and gneiss, which it cuts at a high angle to the foliation. Tuff-breccia of the San Juan is absent in the northern part of the mine. Here the top of the schist is broken and rubbly, and the veins of the system are poor in the schist-rhyolite contact zone.

REPLACEMENT DEPOSITS

Two deposits of the replacement type were studied briefly. One is at the March crosscut of the Osceola mine, just north of the Pride of the West mine, and the other is at the Marcella mine on the western edge

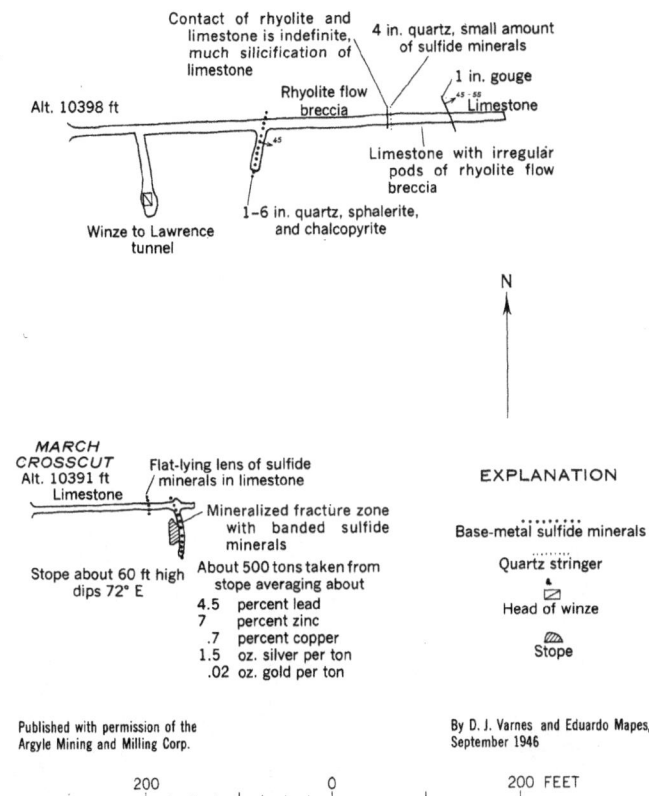

FIGURE 26.—Geologic sketch map of the upper levels of the Osceola mine.

of the map. In both deposits the ore minerals occur in metamorphosed Leadville limestone and Ouray limestone.

The replacement ore bodies of the Osceola mine (fig. 26) occur where the Pride of the West vein zone penetrates large blocks of limestone. Whether these areas of limestone are blocks incorporated in Eureka rhyolite or are in place could not be determined. The ore consists of sphalerite, galena, pyrite, and chalcopyrite in a gangue of quartz and metamorphosed limestone. Much of the ore is peculiarly banded; layers of sulfides a millimeter or two thick, alternate with thin bands of finely crystalline quartz. The ore body in the March crosscut was irregular but apparently lens-shaped, and it followed fractures that trend northward to N. 20° W. and dip steeply eastward. The ore is cut off abruptly against rolls of unmineralized but highly altered country rock.

The Marcella mine (fig. 27) is at the contact between the quartz monzonite and the Leadville and Ouray limestones near the Animas River, south of Silverton. The portal is in quartz monzonite, but most of the winding tunnel is in limestone, marble, and contact-metamorphic rocks. The mineral assemblage present here is of a type characteristic of higher temperatures

than at the Osceola deposit, and apparently was formed through pneumatolytic action along the contact between the quartz monzonite stock and the limestone. Small amounts of galena and sphalerite have been found in the metamorphic zone but not in sufficient quantity to warrant mining. The limestone near the portal has been converted to an aggregate of granular calcite, tremolite, monticellite, diopside, idocrase, garnet, and a mineral tentatively identified as thaumasite. Farther underground, the tunnel veers away from the contact zone and passes through marble and relatively unaltered limestone. Near its end the tunnel approaches the contact again and passes through zones that may indicate progressively higher degrees of metamorphism. The first zone, a few inches wide, consists of tremolite and wollastonite. This is followed by a zone a few inches to a few feet wide, of massive garnet. The sulfides occur sparingly in these two zones. The third zone, which was never completely penetrated, consists of black hornfels composed of fine-grained granular quartz with sericite and magnetite. The mine was examined with an ultraviolet lamp for the tungsten mineral scheelite but none was found. The black hornfels crops out on the hillside above the mine. Between the hornfels and the limestone there are a few small bodies of massive garnet and magnetite.

DISSEMINATED DEPOSITS

The rocks within the ring-fault zone bounding the area of subsidence on the south and west are highly broken, stained, and altered. They form conspicuous red and yellow outcrops along Mineral Creek northwest of Silverton and along the south side of the Animas Valley south and east of Silverton. Along many parts of the ring-fault zone, the Burns quartz latite is almost completely altered to an aggregate of quartz, sericite, chlorite, and clay minerals. The staining is due largely to the decomposition of finely disseminated pyrite. The amount of pyrite and the degree of alteration are greatest where the ring-fault zone is intersected by the stronger northwestward-trending veins.

Mining within the fault zone has been limited to the northwestward-trending veins, such as the Little Nation and Aspen lodes. Very little attention has been given to the possibility of large low-grade deposits within the broken rock of the ring-fault zone. A few grab samples of altered rock assayed from a trace to 0.01 ounce gold and 0.1 to 0.7 ounce silver per ton. These few analyses indicate only that the stained rocks are somewhat mineralized. It is possible, however, that large-scale sampling might delimit areas of lesser and greater mineralization and that some of the areas are worthy of more detailed study.

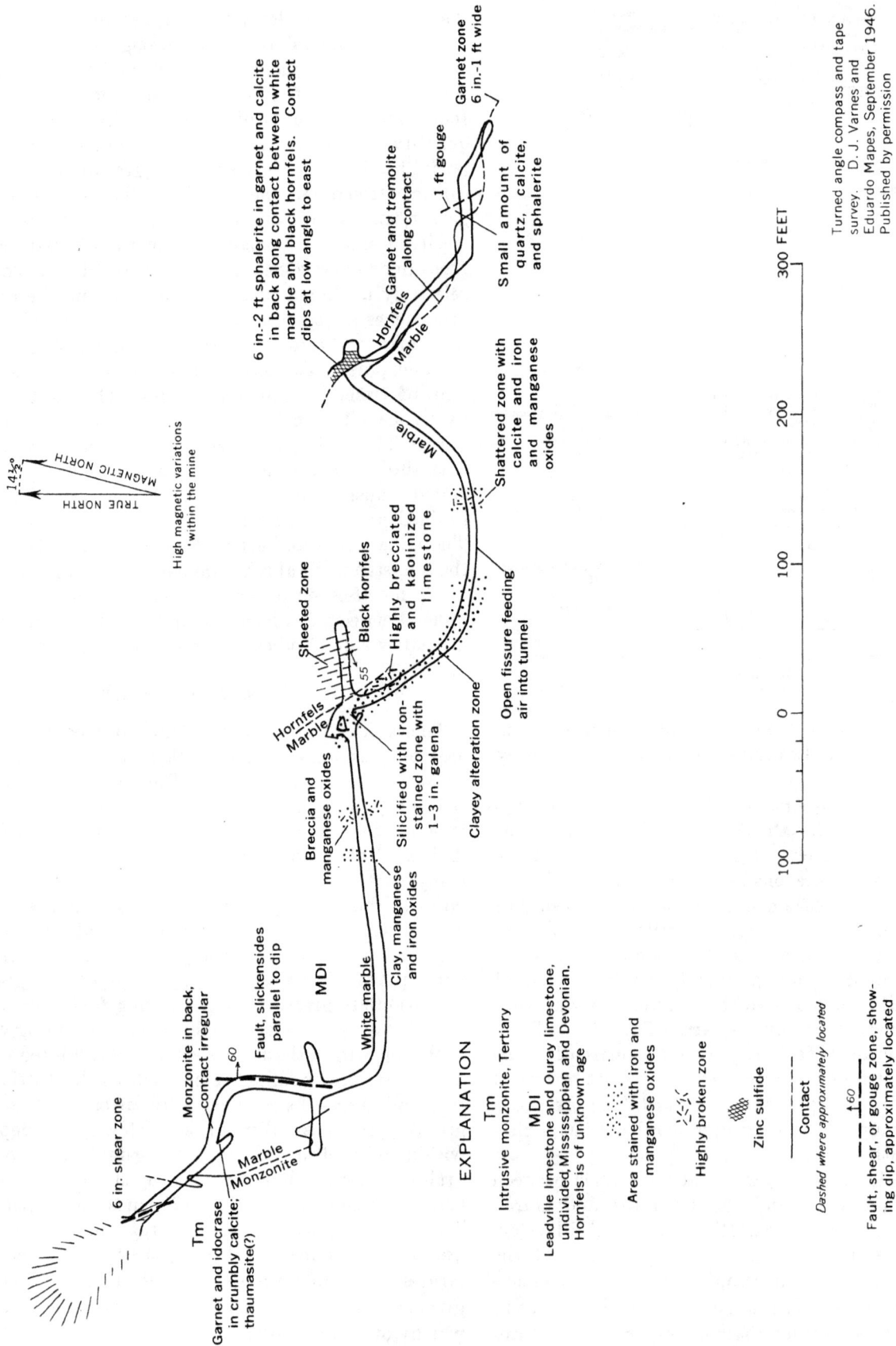

FIGURE 27.—Geologic sketch map of the Marcella mine.

SUGGESTIONS FOR FUTURE PROSPECTING

The main veins in the region to the southeast of Silverton are generally well exposed and have been thoroughly prospected. Future mining appears to depend upon finding new veins that are not exposed, such as those that are "blind," in that they do not extend to the surface, or those whose outcrops are concealed by talus, glacial debris, or by soil and heavy timber.

The possibility of finding extremely rich silver ore is remote, for if the interpretation of zoning is correct, this ore occurred predominantly at the higher altitudes, where it was easily recognized and exploited at an early date. Moreover, the possibility of finding large additional reserves of low-grade ores of gold, silver, and base metals does not seem encouraging. Most future explorations, based on present information, will be highly speculative, or "long-shot" gambles; but there are several types of deposits or areas that might be considered for further study, if this is kept in mind.

1. At the time the mine was visited, there appeared to be a possibility of finding more ore at the Pride of the West mine by exploratory drilling for veins along steep tension fractures between the less steeply dipping faults. This type of prospecting offers more promise of success in the northern than in the southern part of the mine. In the southern part, the faults themselves are mineralized, and steep ore-bearing tension fractures are apparently less common.

2. The Shenandoah-Dives vein system to the northwest of the present workings should be considered; this area lies under talus on the eastern side of Arrastra Gulch. Although the Mayflower vein of the Shenandoah-Dives system is known to continue to the northwest for at least 2,000 feet from the main portal, this vein might be regarded as but one member of a family of veins having a steplike arrangement. The Mayflower vein is on the footwall side of the North Star vein; it is being mined to the northwest of the North Star vein and at a lower altitude. Thus, in tracing the vein system northwestward, it would be most reasonable to explore for another member of the family of veins on the footwall side of the Mayflower vein, as projected at a lower altitude.

3. Veins along tension fractures in the footwall of the Silver Lake vein system are spaced at approximately equal intervals. Deep drilling into the footwall of the Silver Lake vein, across a zone indicated to be most favorable by the spacing of the known fractures, might find another member of this group of rich veins. These diagonal veins may not crop out at the surface. Apparently no exploration, other than on the surface, has been made of the area west of the New York vein, and the New York vein itself has not, at the time of mapping, been explored at the low altitude of the main level crosscut between the Shenandoah-Dives and Silver Lake mines.

4. Some means might be considered for estimating the potential of the northwestern part of the Titusville vein, northwest of the parts mined in the past, and particularly in the vicinity of Kendall No. 2 Peak. Some sulfides are visible at the surface here and there in this area, but apparently there has been little serious exploration.

The structural analysis presented in chapter B does not directly furnish any new clues for finding ore. It does, however, offer some rational basis for placing known veins, and others that may be found, into a coherent structural pattern. The structural analysis should aid in seeking the continuation of a vein underground that has pinched out or that has been terminated by a cross fault.

With reference to pinching out, individual ore shoots in a shear zone may be expected to lay somewhat askew to the general trend of the shear zone as shown in figure 13. They may be close together, even overlapping, or almost parallel with the trend of the shear zone. In any event, when one fracture is followed to its end, the most reasonable direction to strike out in search of the next is determined by the direction of relative movement in the shear zone. If movement was right lateral, as it was in all the major northwestward-trending shear zones, exploration should be to the left, and perhaps ahead, of the fracture that was followed to its end in order to find the next. If movement was left lateral, exploration should be to the right.

The same general rule holds for cross faults in a system of conjugate shears. In exploring along a shear zone having right-lateral displacement that is cut off by another fault oriented at a high angle, 45° to 90°, to the shear zone, one should look for its continuation to the left beyond the cross fault. The rule is reversed if a fault with left-lateral movement is being followed. Owing to local irregularities in the regional pattern or possibly to rotational movements of the fault blocks these rules may not hold invariably, but in cases where other data to guide exploration are lacking they should serve as fairly reliable guides. It is important to recognize what part the local structure plays in the regional fault pattern in order to infer the directions of right- and left-lateral shearing in the absence of direct evidence. The trend of a fault may not everywhere be a reliable indicator of the direction of faulting along it. Some faults, particularly in the eastern shear system, curve through great arcs, but nevertheless preserve their direction of relative movement.

USE OF THE CLAIM MAP

Compilation of the claim map (pl. 7) was undertaken as an adjunct of the geologic mapping in the belief that such a map, on a topographic base, would be useful to the mining community. The adjustment of U.S. Land Office plats to the topographic base proved to be difficult. Few mineral locating monuments and claim corners could be identified positively in the field, although most of the area was traversed on foot and a constant watch kept. In general, the claim map was constructed by identifying features such as portals, shafts, and stream courses that appeared on both the topographic sheet and the plate of a group of claims tied to a particular mineral monument. These features were superposed and then the claim boundaries were transferred to the topographic base. This process was repeated for other groups of claims referenced to other mineral monuments. In some areas large discrepancies between plats and base map had to be distributed gradually between distant tie points. The map is therefore not to be regarded as being of engineering accuracy or as having any legal status, but simply as a guide to the approximate locations of claims relative to their topographic and geologic setting.

During this compilation, I had access to detailed claim maps in the files of the Shenandoah-Dives Mining Company and the Highland Mary mines, and was aided by continued advice from Mr. Joe Vota, then County Engineer.

REFERENCES CITED

Atwood, W. A., and Mather, K. F., 1932, Physiography and Quaternary geology of the San Juan Mountains, Colorado: U.S. Geol. Survey Prof. Paper 166.

Barnes, Harley, 1954, Age and stratigraphic relations of Ignacio quartzite in southwestern Colorado: Am. Assoc. Petroleum Geologists Bull., v. 38, p. 1780–1791.

Burbank, W. S., 1930, Revision of geologic structure and stratigraphy in the Ouray district of Colorado and its bearing on ore deposition: Colorado Sci. Soc. Proc., v. 12, no. 6, p. 151–232.

———— 1933a, Vein systems of the Arrastre Basin and regional geologic structure in the Silverton and Telluride quadrangles, Colorado: Colorado Sci. Soc. Proc., v. 13, no. 5, p. 136–214.

———— 1933b, Epithermal base-metal deposits in Ore Deposits of the Western States (Lindgren Volume), p. 641–652, Am. Inst. Mining Metall. Engineers, New York.

———— 1935, Camps of the San Juan: geologic guides are sought for ore development: Eng. Mining Jour., v. 136, no. 8, p. 386–389, 392.

———— 1940, Structural control of ore deposition in the Uncompahgre district, Ouray County, Colorado: U.S. Geol. Survey Bull. 906–E, p. 109–265.

———— 1941, Structural control of ore deposition in the Red Mountain, Sneffels, and Telluride districts of the San Juan Mountains, Colorado: Colorado Sci. Soc. Proc., v. 14, no. 5, p. 141–261.

———— 1951, The Sunnyside, Ross Basin, and Bonita fault systems, and their associated ore deposits, San Juan County, Colorado: Colorado Sci. Soc. Proc., v. 15, no. 7, p. 285–304.

Burbank, W. S., Eckel, E. B., and Varnes, D. J., 1947, The San Juan region [Colorado]: Colorado Mineral Resources Board Bull., p. 396–446.

Chase, C. A., 1929, A geological gamble in Colorado meets with success: Eng. Mining Jour., v. 128, no. 6, p. 202–205.

Comstock, T. B., 1883, Notes on the geology and mineralogy of San Juan County, Colorado: Am. Inst. Mining Metall. Eng. Trans., v. 11, p. 165–191.

Cross, C. W., 1904, A new Devonian formation in Colorado: Am. Jour. Sci., 5th ser., v. 18, p. 245–252.

Cross, C. W., and Hole, A. D., 1910, Description of the Engineer Mountain quadrangle, Colorado: U.S. Geol. Survey Geol. Atlas, Folio 171.

Cross, C. W., Howe, Ernest, and Irving, J. D., 1907, Description of the Ouray quadrangle, Colorado: U.S. Geol. Survey Geol. Atlas, Folio 153.

Cross, C. W., Howe, Ernest, Irving, J. D., and Emmons, W. H., 1905, Description of the Needle Mountains quadrangle, Colorado: U.S. Geol. Survey Geol. Atlas, Folio 131.

Cross, C. W., Howe, Ernest, and Ransome, F. L., 1905, Description of the Silverton quadrangle, Colorado: U.S. Geol. Survey Geol. Atlas, Folio 120.

Cross, C. W., and Larsen, E. S., 1935, A brief review of the geology of the San Juan region of southwestern Colorado: U.S. Geol. Survey Bull. 843.

Eastman, C. R., 1904, On Upper Devonian fish remains from Colorado: Am. Jour. Sci., 5th ser., v. 18, p. 253–260.

Emmons, S. F., 1888, Structural relations of ore deposits: Am. Inst. Mining Metall. Eng. Trans., v. 16, p. 804–839.

Endlich, F. M., 1876, Report [on the San Juan district, Colorado]: U.S. Geol. and Geog. Survey Terr. (Hayden), Ann. Rept. 8, p. 181–240.

Henderson, C. W., 1926, Mining in Colorado: U.S. Geol. Survey Prof. Paper 138, p. 210–216.

Kelley, V. C., 1946, Geology, ore deposits, and mines of the Mineral Point, Poughkeepsie, and Upper Uncompahgre districts, Ouray, San Juan, Hinsdale Counties, Colorado: Colorado Sci. Soc. Proc., v. 14, no. 7, p. 289–466.

Kirk, Edwin, 1931, The Devonian of Colorado: Am. Jour. Sci., 5th ser., v. 22, p. 222–240.

Larsen, E. S., Jr., and Cross, C. W., 1956, Geology and petrology of the San Juan region, southwestern Colorado: U.S. Geol. Survey Prof. Paper 258.

Nockolds, S. R., 1954, Average chemical compositions of some igneous rocks: Geol. Soc. America Bull., v. 65, p. 1007–1032.

Prosser, W. C., 1914, Silver Lake Basin, Colorado: Eng. Mining Jour., v. 97, no. 25, p. 1229–1231.

Ransome, F. L., 1901, A report on the economic geology of the Silverton quadrangle, Colorado: U.S. Geol. Survey Bull. 182.

Rhodes, F. H. T., and Fisher, J. H., 1957, Ignacio quartzite of southwestern Colorado: Am. Assoc. Petroleum Geologists Bull., v. 41, no. 11, p. 2508–2518.

Richmond, G. M., 1954, Modification of the glacial chronology of the San Juan Mountains, Colorado: Science, v. 119. p. 614–615.

Rickard, T. A., 1903, Across the San Juan Mountains: Eng. and Mining Jour., v. 76.

Riedel, Wolfgang, 1929, Zur mechanik geologischer Brucherscheinungen: Centralbl. Mineralogie, Abt. B, p. 354–368.

Van Horn, F. R., 1901, Andesitic rocks near Silverton, Colorado: Geol. Soc. America Bull., v. 12, p. 4–9.

Varnes, D. J., 1946, Geologic mapping by means of graphic locator: U.S. Geol. Survey Circ. 12, 4 p.

—— 1948, Geology and ore deposits of the South Silverton area, San Juan County, Colorado: Colorado Mining Assoc., 12 p.

—— 1962, Analysis of plastic deformation according to von Mises' theory, with application to the South Silverton area, Colorado: U.S. Geol. Survey Prof. Paper 378–B, p. B1–B49.

Woolf. D. O., 1953, Results of physical tests of road-building aggregate to Jan. 1, 1951: Bur. Public Roads, U.S. Dept. Commerce.

INDEX

O

This original publication came with additional plates and/or maps included in the document or the back pocket of the publication. Miningbooks.com has digitally scanned and formatted these plates and/or maps and put them on a CD ROM. Due to the printing and distribution process we use, some of our distributors and retailers may not have the capability to add this CD in with their drop-ship process. If you did not receive this CD Rom with your book we have made this CD available directly from our website www.miningbooks.com for you to purchase at a nominal cost in order to cover shipping , handling, and materials.

www.ingramcontent.com/pod-product-compliance
Lightning Source LLC
Chambersburg PA
CBHW051233200326
41519CB00025B/7355